IANVARY 1. 1676

London's Leonardo –
The Life and Work of Robert Hooke

London's Leonardo – The Life and Work of Robert Hooke

JIM BENNETT

MICHAEL COOPER

MICHAEL HUNTER

LISA JARDINE

Great Clarendon Street, Oxford OX2 6DP

Oxford University Press is a department of the University of Oxford.
It furthers the University's objective of excellence in research, scholarship, and education by publishing worldwide in

Oxford New York

Auckland Bangkok Buenos Aires Cape Town Chennai Dar es Salaam Delhi Hong Kong Istanbul Karachi Kolkata Kuala Lumpur Madrid Melbourne Mexico City Mumbai Nairobi São Paulo Shanghai Singapore Taipei Tokyo Toronto

Oxford is a registered trade mark of Oxford University Press
in the UK and in certain other countries

Published in the United States
by Oxford University Press Inc., New York

First published 2003

British Library Cataloguing in Publication Data

Data available

Library of Congress Cataloguing in Publication Data

ISBN 0-19-852579-6

Typeset in Plantin
by Graphicraft Ltd., Hong Kong
Printed in Great Britain by
T.J. International Limited,
Padstow, Cornwall

CONTENTS

FIGURES

PREFACE

ROBERT HOOKE (1635–1703) WAS A POLYMATH who has never achieved the recognition he deserves. As Curator of Experiments to the Royal Society, Gresham Professor of Geometry, Surveyor to the City of London, author, and prolific inventor, Hooke at once dazzles us and challenges the boundaries of modern expertise. This book seeks to do justice to his varied achievements by juxtaposing four accounts of the man by specialists approaching him from different but intersecting points of view. Cumulatively, this should give a greater insight into Hooke than would be possible for any study by a single author.

First, Michael Cooper looks at Hooke's childhood and schooldays before presenting an overall view of his career, considering his relations with his various employers and assessing his overall earnings. Then Jim Bennett gives an overview of Hooke's extraordinary fertility in devising scientific and other instruments and devices. Third, Michael Hunter evaluates Hooke's achievement in natural philosophy—a term used throughout the book to describe what might now be called 'science', involving both speculation about the working of the natural world and its experimental investigation. Last, Lisa Jardine assesses Hooke as a man, reconstructing the conflicting pressures he was under, and giving an account of the heroic regime of medication to which he subjected himself.

Although Hooke was brought up in the Isle of Wight and served his apprenticeship in Oxford, by far the bulk of his life was spent in London, and almost all the activities described below are redolent of the metropolitan settings in which they occurred. This explains the title of our book, which was suggested by a paper on Hooke by Edmund Hambly (1987), an enthusiast for Hooke who combined his private practice in civil and offshore engineering with lecturing and writing until his untimely death in 1995. We here pay tribute to the late Dr Hambly, and we also wish to express our gratitude to the Council of Gresham College for inviting the co-authors of this book to give illustrated talks, based on their contributions, at the college in April 2002. In addition, we would like to thank those who have helped in different ways in its preparation: Mordechai Feingold (who read the entire text and made valuable suggestions for improving it), John Henry (whose comments on Michael Hunter's contribution were also invaluable), the library staff of the Royal Society, the archivist and staff of the Corporation of London Records Office, the staff at Guildhall Library, London, the Mercers' Company archivist, and the library staff at City University. The illustrations are reproduced by kind permission of the President and Council of the Royal Society, the Corporation of London, Willen Church Council, Guildhall Library, and the Museum of the History of Science, Oxford.

1

Hooke's Career

MICHAEL COOPER

A Country Child and Westminster Scholar

For more than 200 years after Hooke's death in 1703, the little that was generally known about his childhood came from two members of the Royal Society who knew him only in his later life: John Aubrey, the vagrant virtuoso, antiquarian, and gossip, and Richard Waller, a business-like virtuoso and linguist. They reported what they heard from others, including Hooke, about his earlier years, Aubrey in *Brief Lives*[1] and Waller in his biographical note at the beginning of his edition of Hooke's *Posthumous Works*.[2] They both knew Hooke well, but at different times: Aubrey in the 1670s when Hooke was in his prime, Waller in Hooke's declining years in the late 1680s and 1690s. When Hooke's earlier diary was published in 1935 the editors added a little to what Aubrey and Waller had told us about Hooke's childhood,[3] but within the last ten years research into Hampshire and Isle of Wight records has brought to light many new and interesting details of the Hooke family and their connections with other residents of the island.

Hooke was born on 18 July 1635, the youngest of four children of the Reverend John Hooke, curate of the Church of All Saints at Freshwater in the Isle of Wight, and of Cecily (née Giles). Of the four children, the two eldest were daughters. Hooke's elder brother John was apprenticed to a grocer at Newport and was twice elected mayor of the town, but he committed suicide in 1678. When Hooke's father died in October 1648, he left to his younger son Robert his best-made chest, all his books, and

£40 in cash, to which was added £10 by the will of Hooke's maternal grandmother, Ann Giles.[4] Robert Hooke was not poor, but this £50 was then the whole of his fortune. Although his family thought he might become apprenticed to a watchmaker, it was his artistic abilities that led him to an apprenticeship in London with the portrait painter Peter Lely. The bright but delicate 13-year-old orphan left home to be taught how to add backgrounds and other ornamentation to Lely's portraits of wealthy patrons. Fifty-five years later he died in his rooms in Gresham College, London, as one of the world's first and greatest experimental scientists. Beneath his bed in a locked chest (possibly the one left him by his father) lay his fortune of almost £10,000 in cash—about the size of a successful merchant banker's estate at the time. I shall be concerned with showing how, throughout his career, he increased his legacy 200 times over.

From the time of his childhood Hooke knew that one day he would have to earn a living. He was taught by his father at home in the expectation that his son would follow him into the Church, but lessons were frequently interrupted by headaches and stomach upsets which affected Hooke so much that his family had doubts about whether he would live to maturity. When his father fell ill, all plans to educate him at home were abandoned, and Hooke was left to follow his own inclinations. He soon found that he had an instinctive understanding of the workings of mechanical devices of all kinds. He made mechanical toys, including a fully-rigged model warship which he sailed across the harbour at Freshwater, its guns firing as it went. After examining the components of a brass clock, he made a working model out of wood. John Aubrey records that when the artist John Hoskins visited Freshwater, Hooke watched what he was doing and then drew some pictures using artists' materials he had made himself from coal, ruddle, and chalk. Hoskins said that the young boy had a natural aptitude for painting.[5]

When using his mind, hands, and eyes in studying and making mechanical contraptions he was able to overcome headaches and sickness. He used his powers of observation in order to understand how things worked so that he could make practical use of that understanding. These abilities were being developed apparently without any instruction from anyone apart from Hooke himself. He was his own guide, using the evidence of his senses to judge whether or not the things he was planning and making were following the right lines. If the ship or clock he designed and made failed, he knew it was because he had insufficient understanding, or inadequate skill in designing and making (or both). In order to succeed he would have to find ways of understanding more and becoming more skilful through his own efforts. In these early years he developed his capacities for observation and mechanical invention, but he told

Richard Waller that he became hunch-backed from the age of 16 through spending too much time bending over a lathe. He was nevertheless sprightly and quick in his movements.[6] His imagination too was sprightly and quick.

Hooke did not stay long in Lely's studio. Waller records that Hooke left because his poor health was aggravated by the smells of artists' materials.[7] On the other hand, Aubrey wrote that Hooke very quickly understood what was required of him in Lely's studio and decided that he should not waste his legacy on an apprenticeship.[8] There is probably some truth in both assertions. The particular circumstances that led Hooke first to Lely and then to Dr Richard Busby's house at Westminster School have only been guessed at, but recently the discovery of some Isle of Wight Anglican and royalist connections has thrown some light on them. Hooke's father was a loyal high Anglican who had tutored the son of Sir John Oglander, a prominent resident of the Isle of Wight. When Charles I fled to Carisbrooke Castle in 1647, the ardent royalist Oglander was at the centre of what was a sort of court in exile, which probably included Hooke's father. John Hoskins, who was so impressed by Hooke's talent for drawing, was Charles I's miniaturist. A witness of the will of Hooke's father was Cardell Goodman, 'a worthy and well beloved friend',[9] vicar at Freshwater and another Anglican royalist, who had been to Westminster School and Christ Church, Oxford. It is reasonable to suggest that Hoskins put Hooke in touch with Lely, who had been in London since 1643 and was patronized by the King, and that Goodman was the link between Hooke and Busby.[10]

Hooke's £50 legacy might have been just enough to cover the fees and other costs of his time at Westminster School.[11] Other evidence, from Busby's account book, shows that pupils were not charged according to set scales, and that for some, the costs could have been as high as £30 a year.[12, 13] On the other hand, Busby would take in and teach a pupil gratis if he thought the boy had special promise. Hooke might well have been allowed to keep his £50. Much later, Busby and Hooke became friends and often dined together. Hooke was the architect for the church and vicarage at Willen in Buckinghamshire (Fig. 1), built in the 1680s at Busby's expense. At Westminster School Hooke learned Latin, Greek, a little Hebrew, quickly mastered the first six books of Euclid, and was competent at musical exercises at the organ. But there is evidence that he also learned from extra-curricular activities. John Aubrey records that one of Hooke's contemporaries at Westminster School noticed that he was not often seen there at lessons.[14] Perhaps Hooke was out and about in the 'dark shops of Mechanicks'[15] watching and talking to craftsmen, more concerned with trying to make the flying machines he

Fig. 1 Church of St Mary Magdalene, Willen, near Milton Keynes (designed by Hooke, 1680s) and the Hospice of Our Lady and St John. The original cupola on the church tower has been removed. (Willen Church Council.)

had in mind than taking formal lessons in Euclid, music, and languages, mastery of which seems to have taken little of his time. If Hooke was right in recollecting that he began to grow 'awry' around the age of 16 through too much bending over a lathe, he would then have been a pupil at Westminster School. A boy's absence from lessons would not have escaped Busby's attention, but he would probably have condoned them if he could see that Hooke's unconventional interests and abilities were being nurtured elsewhere without detriment to more formal academic studies.

A Sort of Apprenticeship

From the time of Queen Elizabeth I a regular procession of pupils left Westminster School for Christ Church, Oxford, including, in 1599, Edmund Gunter, who became Gresham Professor of Astronomy in 1620, and the physician and philosopher John Locke, who went to Christ Church in 1652, the year before Hooke. Unlike Gunter and Locke, who were elected, Hooke was awarded a Choral Scholarship which was 'a pretty good allowance', according to John Aubrey.[16] Richard Waller tells us that Hooke also received benefits at this time as Servitor to a Mr Goodman. No student named Goodman appears to have been at Christ Church during Hooke's time, so probably Cardell Goodman, the 'worthy and well beloved friend' of Hooke's father, provided for Hooke at Christ Church. Hooke matriculated in July 1658 and was admitted MA in 1663.

John Wilkins, Warden at Wadham College when Hooke arrived in Oxford and later Bishop of Chester (Fig. 2), already knew of Hooke's precocious mechanical ingenuity. When Hooke was still at Westminster School, Wilkins had sent him as a gift a copy of his *Mathematical Magic* (published in 1648) which, with its emphasis on the usefulness of mechanical devices and demonstrations of how 'a divine power and wisdom might be discerned, even in those common arts which are so much despised',[17] gave Hooke an incentive to practice his innate mechanical gift. It was to Wilkins that Hooke showed his flying devices at Oxford and it was Wilkins who received fulsome praise from Hooke in the preface to *Micrographia* as the man who encouraged him to follow Wilkins's protégé Christopher Wren in making microscopical observations: 'there is scarce any one invention, which this Nation has produc'd in our Age, but it has some way or other been set forward by his [Wilkins's] assistance. . . . He is indeed a man born for the good of mankind, and for the honour of his Country.'[18] Wilkins more than anyone was Hooke's inspiration in natural philosophy and the source of his optimism.[19] Wilkins convinced Hooke that mechanics was a fit subject for intellectual study as well as practice, and that through thoughtful and skilled application it would bring benefits to mankind.

Fig. 2 John Wilkins when Bishop of Chester. Engraving by A. Blooteling from a painting by M. Beale. (Private collection.)

Another man at Oxford who had an early influence on Hooke was the Savilian Professor of Astronomy, Seth Ward.[20] At his suggestion, Hooke devised a way of improving the motion of the pendulum for timing astronomical observations.[21] Hooke was making a reputation for himself at Oxford, so it is not surprising that he was invited to work in the laboratory of the physician, chemist, and loyalist Anglican churchman Thomas Willis.[22] Hooke lodged in Willis's house in Merton Street, opposite Merton College chapel. Busby had been the first of a succession of men in positions of authority who noticed that Hooke was exceptionally able and ambitious. Willis was the first of many who benefited from seeing that Hooke's special abilities could lead to personal, institutional, or public benefit. He contributed to Willis's two main activities: his practice as a physician and his academic interest in iatrochemistry—the use of chemical elements and compounds for treating illness. Hooke not only learned about such topics directly from Willis, but through experience he increased his understanding of the properties of different laboratory materials and developed his ability to measure, shape, and join components together for specific purposes. Skills such as these were not unusual; they could be found among master craftsmen, but Hooke was not apprenticed. He possessed abilities, rarer and more valuable than his high technical competence—an inquiring mind and a fertile imagination. All these attributes made him well suited for a role in natural philosophy.

He was soon taken up and sponsored by another gifted man who gave him the opportunity to develop further his technical and experimental expertise, and so to reach a position where he had an income for most of his life and through which he became one of the greatest experimental scientists of his time.

The Right Honourable Robert Boyle (see Fig. 52), the fourteenth child and seventh son of the Earl of Cork, pursued his scientific investigations as an aristocratic and Christian gentleman of private means, living on an estate at Stalbridge in Dorset inherited from his father. In the winter of 1655/6 he moved to Oxford, but set up his own private laboratory in a property rented from the apothecary John Crosse. Known as 'Deep Hall', it lay on the south side of High Street adjoining the west side of University College. Boyle worked and lived there until 1668, when he moved to London to share the house of his sister Lady Ranelagh in Pall Mall. There for the rest of his life he continued to work and receive visitors to his laboratory from home and abroad.

During the 1650s, Boyle became interested in the physical properties of air, a substance known to be essential for life and therefore worthy of scientific investigation. At the same time in Magdeburg, Otto von Guericke was using an air-pump to show that air had weight: a vessel emptied of air weighed less than when it was full. He also demonstrated the existence of atmospheric pressure and the huge effort needed to overcome it by showing that two teams of six strong men could not pull apart two copper hemispheres of 200 mm diameter placed together, from which air had been pumped out. Though von Guericke's experiments were not published in full until 1672, Boyle heard about them through the account published by Gaspar Schott in 1657, and he determined to replicate them. At first he had an air-pump made for him by Ralph Greatorex, a mathematical instrument-maker, but it was poorly constructed from unsuitable materials and could not prevent air leaking into the evacuation chamber. Hooke described it as 'too gross to perform any great matter'.[23] Boyle asked Hooke to design and make a new pump, so he went to London and made (or gave instructions for making) a new cylinder barrel from better materials. The new air-pump made by Hooke on his return to Oxford was a remarkable achievement in itself, but it was more important in that it provided Boyle, Hooke, and others during the next few years (until it wore out and had to be replaced) with the means of making qualitative and quantitative experiments (see Fig. 51).

Together Boyle and Hooke performed forty-three experiments with the new air-pump, which were reported in Boyle's first scientific publication, *New Experiments Physico-Mechanical, Touching the Spring of the Air and its Effects*, published in Oxford in 1660. Subsequently, in 1661, Boyle and Hooke's further experiments revealed the relationship between the pressure and the volume of air which is now known as 'Boyle's Law'—the volume of a gas at constant temperature is inversely proportional to its pressure. Measured data showing that the observed variation between pressure and volume is reciprocal were published in 1662 in Boyle's *Defence against Linus*, an appendage to the second edition of *New Experiments*. Independent experiments and demonstrations were later performed at the Royal Society and elsewhere by Hooke and others before the hypothesis was generally accepted. Its revelation through experiment and measured data is typical of many scientific discoveries, from Galileo's experiments on falling bodies about sixty years before Boyle's Law to those of the present day.

Although Boyle employed various men to assist him in his experiments by making and using apparatus under his instructions, in working with a brilliant, but socially much inferior, young man like Hooke, both men entered a professional relationship in which modes of behaviour had to be worked out and mutual trust established. Hooke was probably appreciative of Boyle's trust in him when he was given a free hand to design and make the air-pump. The great success he made of that opportunity was acknowledged by Boyle in the first edition of *New Experiments*, where he names Hooke as the maker of the successful apparatus. Steven Shapin has shown that Boyle's technicians were generally anonymous and their contributions unacknowledged in published accounts of the experiments in which they took part. To have made them visible would generally have compromised their employer's authority and integrity.[24] In mentioning Hooke by name, Boyle showed that he regarded him as very different from his usual technicians. The extent of the difference is shown by Edward B. Davis in his examination of an incident in which Boyle's trust in Hooke is shown to extend beyond technical and scientific matters to giving advice on Boyle's strategy in dealing with his philosophical disagreements with Descartes and others.[25] Hooke contributed much more to Boyle's work than might be expected from even an exceptionally ingenious technician. Boyle's Law is an early demonstration of the fact that cooperative scientific investigations can be undertaken successfully despite wide differences in the social backgrounds and personal characteristics of the people involved in them. Boyle's inherited wealth and Hooke's practical expertise together revealed Boyle's Law, but neither wealth nor practical expertise alone was sufficient. It was necessary also to reconcile two quite different creative imaginations in reaching agreement on what to do and why, who should do it, and who should report it. Hooke and Boyle thought and worked together in new ways. How they achieved this meeting of minds can only be a matter of speculation. In any case, the outcome was an important early example of success in experimental science and it completed a sort of apprenticeship for Hooke.

Curator of Experiments and Cutlerian Lecturer

The origins of the Royal Society are usually thought to go back to around 1645, when various physicians, scholars, and clerics, all of whom had an interest in natural philosophy, began to hold informal meetings in London to make observations of natural phenomena and debate what they had seen. Subsequently, some of these men moved to Oxford, forming part of the scientific grouping around John Wilkins at Wadham College with which Hooke was associated. The Restoration brought a number of these men to London, along with others who had been associated with the exiled Stuart court, and at this point meetings took place at Gresham College. On 28 November 1660, this group of men, gathered in the rooms of Laurence Rooke, Gresham Professor of Geometry, decided to regularize their activities by forming a Society to pursue natural philosophy at weekly meetings by experimenting, debating, and recording what was said and done. They appointed officers and obtained rooms in Gresham College for a library and a repository, and a larger room where they could gather every Wednesday afternoon at 3 p.m. for their meetings.[26] They decided to appoint two servants: an amanuensis (or clerk) at an annual salary of £4;[27] and an operator (or technician) at an annual salary of £4, to make and repair equipment to be used in experiments. They also decided that all costs and other expenses should be borne by income from the membership as a whole through admission fees and subscriptions. The Society's apparently innocuous fiscal arrangements set a pattern that, according to Michael Hunter, led to 'debilitating tension between its ideals and its actual state in its early years'.[28]

In forming an independent institution for science, the Society was setting itself a task that was ambitious and unprecedented. To increase income, more members were admitted, but too many failed to pay their dues.[29] Through its first ten years the Society's arrears steadily increased to almost £1,500. Lack of income was a continuing source of anxiety. In the expectation that a subvention from the Crown might be forthcoming, the Society sought and obtained the King's approval of their aims, but royal patronage and incorporation as 'The Royal Society' in 1662 did not immediately bring in funds from the monarch. On the contrary, when the court heard about the Society's early investigations into the properties of the atmosphere, some of the courtiers were highly amused that men supposedly so clever should be so silly as to pass their time weighing air. Furthermore, membership of the Society was not confined to those with an active interest and ability in experimental philosophy—they were too few—so admittance to the Society increasingly became based on the social standing and personal wealth of individuals. The Secretary, Henry Oldenburg, wrote in a letter to Boyle in 1664:

> we grow more remiss and carelesse . . . our meetings are very thin . . . [the Society could prove] a mighty and important Body . . . if all the members thereof could but be induced to contribute every one their part and talent for the growth, and health and wellfare of their owne body.[30]

Experiments were an essential part of the Society's activities. In 1661 and 1662 various members were appointed as ad hoc curators to perform experiments at the Society's weekly meetings. The amanuensis was sometimes ordered to make equipment and even perform experiments in addition to the clerical duties for which he was employed, but he could not make even the simplest apparatus without careful and detailed instructions on how to proceed. Few members were capable of giving useful technical instructions. Even fewer had the skills to make the equipment, an unseemly activity for a gentleman. The proposed annual salary of £4 for an operator was quite inadequate to secure anyone capable of making unaided the new apparatus and instruments needed for experiments. The prospect of realizing the idea of an institution for natural philosophy—the 'Solomon's House' proposed by Francis Bacon in his posthumous *New Atlantis* (1627)—was becoming more distant with each meeting. The Society was discovering that scribes and craftsmen could not be transformed into productive philosophical servants. Members were becoming increasingly frustrated by the lack of progress. They needed someone with the skills to make new kinds of instruments and perform experiments to order.

Hooke was not one of the formative members of the Society, but he was well known to many through his work in Oxford for Thomas Willis and Robert Boyle. His name is mentioned for the first time in Society records on 10 April 1661, when members decided to debate at their next meeting his recent publication in Oxford of a tract on capillary attraction, *Explication of the Phænomena*, but he was not present.[31] The Society was excited by the quality of the experiments underlying Boyle's *New Experiments* and Hooke's *Explication of the Phænomena*, but its members were unable to design and carry out such experiments themselves. One of the leading members of the Society, Sir Robert Moray (who had been active in restoring Charles II to the throne and had used his influence to gain the King's approval of the new Society and its incorporation in 1662) decided that the custom of nominating members as ad hoc curators had not been successful and that a regular curator of the Society's experiments was clearly necessary. On 5 November 1662 he proposed that the Society should find

> a person willing to be employed as a curator by the society, and offering to furnish them every day, on which they met, with three or four considerable experiments, and expecting no recompense till the society should get a stock enabling them to give it. The proposition was received unanimously, Mr. Robert Hooke being named to be the person.[32]

They had recently seen at first hand evidence of Hooke's ingenious workmanship in the air-pump which Boyle had presented to the Society and which had already worked successfully in some of their demonstrations. Earlier in the year Hooke had also attended some sea trials of pendulum clocks with members of the Society. He was an obvious choice for employment as Curator of Experiments, but he was working in Oxford with Boyle and correct procedures had to be followed.

At the Society's next meeting on 12 November 1662, Moray formally proposed Hooke as Curator of Experiments. With Boyle's agreement, Hooke eagerly accepted the opportunity to follow an experimental career. Although no remuneration from the Society was forthcoming, it is likely that some informal arrangements were agreed between Boyle and the Society to share Hooke's services, with Boyle continuing to employ him until the Society could afford a curator's salary. There was unanimous agreement to the appointment. The Society

> ordered that Mr Boyle should have the thanks of the society for dispensing with him for their use; and that Mr. Hooke should come and sit amongst them, and both bring in every day of the meeting three or four experiments of his own, and take care of such others, as should be mentioned to him by the society.[33]

The apparent simplicity of this arrangement concealed difficulties that would cause Hooke and the Society continuing frustration. He was not only to devise and perform three or four of his own experiments, but also any others thought of by members—perhaps as many as six at each meeting. This was a greatly over-ambitious programme. The Society needed him not only for his mechanical ingenuity, but also for his capacity for philosophical speculation, which they had admired in *Explication of the Phænomena*. At a time when institutional procedures for performing and reporting scientific investigations were being worked out, the profession of an experimental scientist was unrecognized. We have seen how Hooke and Boyle had already found a way for disparate individuals to carry out cooperative experiments successfully and write about them with authority, but the Society took much longer to find broad procedures for successful institutional science. One of the difficulties was that they had yet to recognize the differences between a learned society for the advancement of natural philosophy and a research institute for making those advances. The Society, naturally enough, began by attempting to be both.

For more than two years Hooke performed his weekly experiments for the Society without receiving any salary. Now a Fellow, he was exempt from all charges placed on the membership as a whole. For reasons of loyalty and necessity he was working part-time for Boyle in Oxford; when he was in London he was provided with lodgings at a house in Pall Mall belonging to Boyle's sister, Lady Ranelagh, and went from there to the Society's meetings in Gresham College. Too many members of the Society refused to pay their fees and subscriptions. It was difficult to see where money to pay Hooke a regular salary might come from. He was playing an increasingly important part in their meetings and was preparing experiments to impress the King in the hope of a subvention or grant to the Society. For two years they looked for an opportunity for him to take up residence in Gresham College. In 1664 a chance came when the Society supported him in his application for the position of Gresham Professor of Geometry. The application failed, but he successfully appealed against the decision. When Sir John Cutler (Fig. 3) a wealthy grocer, offered to pay Hooke £50 annually to give lectures on the history of trades in his (Cutler's) name at Gresham College throughout Hooke's lifetime, the Society saw the opportunity it was looking for. On 22 June 1664 the Council of the Royal Society decided to investigate how to secure Cutler's offer.

Fig. 3 Statue of Sir John Cutler from the façade of the College of Physicians in Warwick Lane; now in Guildhall. (Guildhall Library.)

A month later, Council voted that Hooke should receive £80 annually (a decent salary) as their Curator of Experiments, payable from subscriptions of various members or otherwise. Council also decided that Hooke should provide himself with lodgings in Gresham College and that these decisions should remain secret 'till Sir John Cutler have Established Mr Hook as a Professor of the Histories of Trades'.[34] The decision implies that the Council was expecting Cutler to establish a kind of Gresham professorship to add to the seven already founded by Sir Thomas Gresham's will.[35] With Hooke receiving a salary from Cutler and living in the building where the Society held its meetings, instead of in Pall Mall, he would find it easier to prepare and perform his experiments. The Council members expected that £50 of the £80 curator's salary they had just approved would come from Cutler for Hooke's lifetime, and they were hoping the remainder would be provided by members' special contributions, at least for the present. Council ordered Hooke to prepare an oration on Sir John Cutler's account (thereby inaugurating the Cutlerian Lectures) and submit, for their approval, details of how he intended to proceed with his lecture programme. The Society was acting as if it, not Hooke, was the recipient of Cutler's patronage, even though public lectures had never been one of its objectives and despite Cutler 'intending a particular kindness to Mr Hooke'.[36] Hooke was caught in a net of contradictions, misunderstandings, and inconsistencies, tangled further by a signed bond between the Society and Cutler, to which Hooke was not party, but which would take years to unravel. As a consequence Hooke endured many years of increasing and debilitating anxiety over personal, legal, and monetary matters that were finally relieved by a court decision made on his sixty-first birthday, which awarded him the outstanding arrears of his salary from Cutler's estate. As Michael Hunter has clearly revealed in his unravelling of the many strands of the Cutler/Hooke/Royal Society disputes,[37] each of the parties had contributed their own particular twist to them.

On 5 October 1664, when Council ordered Hooke to give his inaugural Cutlerian Lecture, he had received no payment of salary from the Society. After more prevarication and indecision, the office of Curator was formally established and Hooke was elected Curator by Office on 11 January 1665 at a salary of £30 per annum *pro tempore*. The decision made official the Society's intention that Hooke's annual curator's salary of £80 should be reduced to £30 because he would be paid £50 annually by Sir John Cutler. Despite this decision, by November 1666, after working for the Society for more than four years, Hooke had received salary payments from the Society amounting to only £50.[38] By then he was living in Gresham College as Professor of Geometry, receiving regular payments of his £50 Gresham salary, notionally receiving another £50 annually from Sir John Cutler, and he was about to be appointed Surveyor to the City of London in the aftermath of the Great Fire. This post would carry important civic responsibilities accompanied by regular and prompt payment of a salary (several times greater than his £30 from the Royal Society) and, in addition, the receipt of innumerable fees from London's citizens. Council knew that his Royal Society salary was much in arrears. It was also clear that his regular participation in their activities as Curator of Experiments was crucial. Council therefore decided to investigate the Society's records to discover what had been done about his salary and what had already been paid to him. Although the details were complicated by uncertainty about the exact agreement with Sir John Cutler, the fact that the investigation took four months to complete shows that Council was somewhat dilatory in respect of Hooke's interests. It was eventually decided that salary arrears of £45 3s. 4d. to Christmas Day 1666 were due.[39] These arrears were paid on 11 November 1667, almost a year late.[40] The Society continued to pay Hooke at an annual salary of £30, but the payments were irregular and often late. By the end of 1675, however, all arrears had been cleared. Hooke then received regular and prompt payments of either £15 every six months or £7 10s. every quarter up to Lady Day 1684.[41]

The Society's bargaining with Sir John Cutler left Hooke in a difficult position. It can be argued that the Society's overall intention was worthy (to raise money to pay a lifetime's salary to their Curator) but through its proceedings Hooke became increasingly frustrated. Cutler refused to pay Hooke his salary as he came to realize that the Society and Hooke had moved from the history of trades to natural philosophy as the subject of the lectures. Nevertheless, Hooke devoted a great amount of time and much intellectual and physical energy to the Society's aims. It is hard to see how the Society could have survived its early years without Hooke's vibrant presence at meetings and the hours he worked at night, sometimes in his workshop with the help of the Society's operator, striving to make new instruments more accurately than before, using his telescopes (Figs. 25, 31) and quadrants for observing stars, planets, comets, and the moon, and his microscope (Fig. 30) for examining the details of things nearer to hand. The results of his painstaking observing and recording at the microscope delight and surprise even now, but to produce his drawings he had to overcome problems of illumination and the aberrations and limitations on depth and area of the field given by his lenses (Fig. 4). Jim Bennett and Michael Hunter discuss below Hooke's purpose and achievements in making and using instruments, and the speculations that arose in consequence.

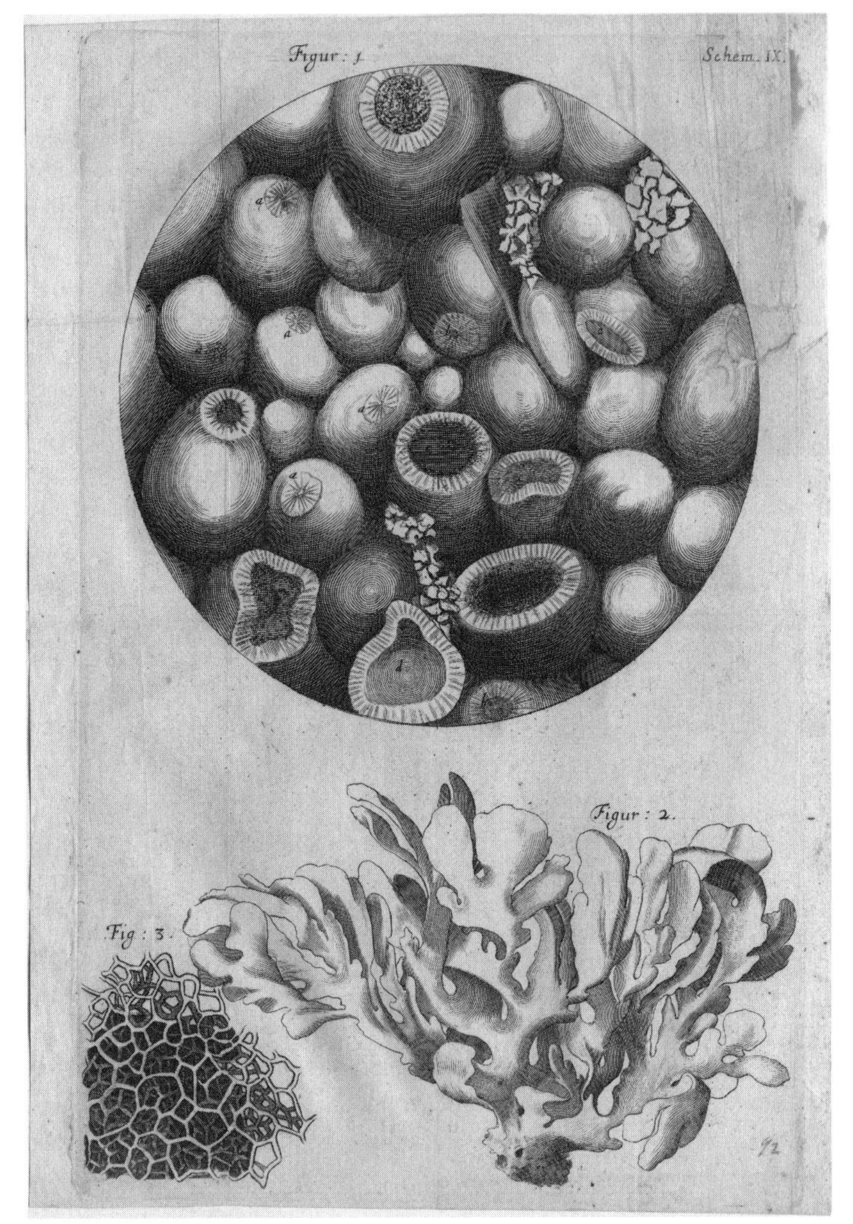

Fig. 4 Hooke's illustration of a fracture surface of 'Kettering-stone' (now known as 'Ketton Stone') taken from a quarry near Kettering in Northamptonshire. Observation XV, Figure 1 of Schema IX in *Micrographia* (1665). The diameter of the original engraving is about 160mm. The actual size of the grains is about 0.5mm. (Museum of the History of Science, Oxford.)

Steven Shapin has shown the importance of Hooke's private rooms in Gresham College (Fig. 5) as a place where he lived and worked in the company of his domestic servants and technical assistants, but which was rarely visited by his colleagues in the Royal Society.[42] Rob Iliffe has shown how in the 1670s Hooke moved across social boundaries and went to places where the exchange of information, skills, and techniques was as much part of an economy as money.[43] Gresham College was such a place, where the separation of the hidden and frequently arduous preparation by Hooke of an experiment from its subsequent performance in front of the Society was similar to the circumstances in which many other events in the lives of some members were organized for them. In science, however, such a separation inhibits the transmission of understanding of the design and technological processes that underlie experimental investigations and which influence their outcome. Hooke's genius and value to the Society can be characterized by his ability on the one hand to design, make, and use a microscope and on the other to discuss, to Boyle's profit, Euclid's *Elements* and the philosophy of Descartes.

Although the Society was short of money, it was beset by another, more fundamental difficulty in connection with Hooke's experimental programme: the excessively high level of expectation that many members of the Society had for experimental investigations. Their enthusiasm for witnessing and discussing experiments was unabated throughout the 1660s. When reading Thomas Birch's published transcripts of the Society's journals at this time,[44] the effervescent spirits and fertile imaginations of Hooke and some other Fellows of the Society emerge from the unadorned prose. But exuberance and imagination were not rewarded by immediate success. The Society became impatient when an experiment 'failed'—by which they often meant that the outcome was not what they expected it to be. But such failures are important in science because in time they bring about rethinking, new lines of enquiry, and improved equipment. In the Society's headlong engagement in experiments and demonstrations intended to gain unprecedented understanding of the natural world, from which great personal and public benefits would flow, patience and concentrated effort were discounted. All the hard work behind their experiments was out of view. It is not possible to discover a Boyle's Law or a Hooke's Law each week. The following list of tasks Hooke was given to perform in just three weeks in October 1663 shows that there were many opportunities for failure, but it also illustrates the Society's thirst for knowledge and its benefits: prepare a paper on what should be observed and recorded for a history of weather; make and demonstrate a hygroscope from the beard of a wild oat, with an index; prepare two thermometers invented by Christopher Wren, one of tin, the other of glass; make an artificial eye; arrange for a suitable concave glass to be made and use it for projecting a picture in a lighted room; cut out a piece of dog's skin and stitch it on again to see if it will grow; take lodgings in Gresham College and supervise the operator in making a new air-pump and a machine for

Fig. 5 Part of George Vertue's 1739 engraving of Gresham College (Fig. 6, p. 22) showing Hooke's rooms (9) in the south-east corner of the quadrangle. Above, on the roof, are the frames of the housing for his zenith telescope. The building with the steeply pitched roof (4) is the College Reading Hall where Hooke gave his Gresham and Cutlerian Lectures. The gate into Bishopsgate street is at (1). (Guildhall Library.)

measuring the force of gunpowder; show microscopical observations of a common fly and of moss growing on a brick; take care of the Society's Repository in the west gallery of Gresham College and place a label on each object so people can know what it is and its provenance; get ready to demonstrate to the King his (Hooke's) new device for taking soundings at sea without using a line; and graft feathers onto a cock's comb.[45]

To the list above can be added a similar number of equally diverse experiments carried out by other members in the same period. Hooke's experimental topics appear random, but they could now be classified under only four headings: the earth and its atmosphere; optics; military and naval technology; and biology. Such a classification was irrelevant at the time, when all the Society's experiments were part of the same new learning. These activities were not so much random as impractically over-ambitious relative to the financial resources, materials, and techniques available, and the way the work was organized. Hooke pursued three different kinds of activities throughout the years of weekly performances demanded by the Society: carrying out experiments designed to lead to understanding of natural phenomena such as the earth's gravity and atmosphere; designing and making useful optical and mechanical devices; and making speculative enquiries upon causes of phenomena and the nature of matter. Far from being independent of one another, each activity benefited from his engagement in the others, as Jim Bennett and Michael Hunter discuss below. What Hooke's instruments enabled him to measure or see provided grounds for speculation which in turn led to new experiments, or the repetition of earlier ones, improved instruments, and new lines of enquiry.

The Society, realizing that Hooke's curatorial duties were excessive, chose the physicians Walter Needham and Richard Lower to relieve him from anatomical tasks, such as the dissection of an embryonic dog he was currently charged with.[46] Hooke was not alone at this time in failing to perform the experiments ordered. At the same meeting, Dr King was absent when he had been ordered to perform an experiment on a dog's thorax—a task he repeatedly failed to complete. Hooke had known Richard Lower at Oxford as a fellow pupil of Thomas Willis and now favoured him as curator of anatomical experiments. On 5 September 1667 Hooke wrote in a letter to Boyle (who also knew Richard Lower at Oxford in the 1650s): 'I hope I shall prevail upon Dr Lower, and for him, so as to get him anatomical curator to the society. He has most incomparable discoveries by him on that subject, and a most dextrous hand in dissecting.' In a postscript, Hooke adds 'Many other things I long to be at, but I do extremely want time.'[47] Despite Hooke's and John Wilkins's attempts to persuade Richard Lower to take up the post of anatomical curator, he refused, preferring his physician's practice to philosophical investigations. He took no part in the Society's affairs after 1668 and was expelled in 1675.[48] In contrast to many members of the Society with social cachet and inherited wealth, Lower had scientific ability, but he could earn a much better living from his practice as a physician than as the Royal Society's anatomical curator.

The Society tried intermittently to provide curatorial assistance for Hooke. On 30 May 1668 he told Council that he had found a man suitable to assist him in his experiments and said, if they employed him at an annual salary of £20, then he (Hooke) would not fail to perform three experiments at every meeting. Council agreed to pay the man pro rata for three months to assist not only Hooke, but all members in their experiments when called upon to do so. This was not what Hooke had intended. No appointment was made.[49] Hooke continued to try to meet his obligations but, lacking regular support, he failed from time to time to do what was ordered.[50] In particular, on 17 June 1669 he received a warning:

> Mr Hooke excused himself for having prepared no experiments for this meeting. He was ordered to take care, that against the next either his own new instrument for working elliptical glasses, or that of Dr. Wren for grinding hyperbolical ones, might be ready; as also that a couple of long pendulums, to be moved by the force of a pocket watch, be prepared, to see how long they would go even together.[51]

The satisfactory completion of tasks such as these needed much more care and attention than Hooke was allowed time for.

Financial hardship and flaws in the Royal Society's structure and procedures began to be a matter for concern in the early 1670s. Plans to reorganize the Society's publications and a proposal to appoint a few committees, each one to oversee a programme of experiments in a specific area, show that the Society was coming to see a difference between promoting and doing science. Michael Hunter shows that Hooke and Henry Oldenburg estimated the annual cost of running a more efficient Society and the value of an investment necessary to secure an income to finance it.[52] At a meeting in February 1671 an anonymous member observed that 'very many things were begun at the society, but very few of them prosecuted,'[53] but no effective action was taken. As the formative members died, or became less active, so the Society relied even more on Hooke for practical demonstrations, but the standard of debate at its meetings declined. In 1673 Seth Ward wrote that the failing or continuance of the Society was brought to a crisis.[54] During this critical time,

Hooke was by far its most productive member, despite his other obligations to Cutler, Gresham College, and the City of London.

Following the death in 1677 of the Society's first Secretary, Henry Oldenburg, Hooke was elected for the first time to Council. He was also elected as one of two Secretaries.[55] His position was now even more complex and his relations with the Society worsened accordingly. As Curator he was given orders by Council and supervised by a Secretary; as a Council member he was collectively responsible for the orders and as Secretary he supervised the Society's salaried employees. As Curator he worked in his rooms with the Society's operator and at his wooden bench in front of the Society, performing experiments at its weekly meetings; as Secretary he sat with other Council officers behind the mace at the cloth-covered table during the weekly meetings, taking notes of the proceedings.[56] Now in his mid-forties, Hooke had not yet lost his eagerness to take on new responsibilities, but he was no longer able to draw on the astonishing resources of physical and mental energy that had earlier enabled him to overcome his generally poor health and meet all his onerous obligations to his various employers. As Lisa Jardine shows below, his use of drugs for medicinal and experimental purposes had a progressive effect on his health and in consequence on his ability to continue to work at the phenomenal rate he achieved in the decade 1665–75.

One of the essential acts in natural philosophy had been neglected by the Society: its members had not kept a detailed and orderly record of all the experiments they had witnessed and of the outcomes, conjectures, hypotheses, and other trials and experiments that had arisen in consequence. Jim Bennett shows below that in *Micrographia*, published in 1665, Hooke wrote that human memory, senses, and reason, being imperfect, needed improvement.[57] The Royal Society had not paid sufficient attention to improving its collective memory. On 7 August 1679, the Council (of which Hooke was a member)

> ordered and desired, that Mr. Hooke do, as soon as may be, print a relation of all the experiments, observations, and relations made and brought into the Society by himself since his first coming into it; and that he hath leave to take his own method in the doing thereof.[58]

This order was at the same time necessary and impossible to meet. By then Hooke had performed literally hundreds of experiments and demonstrations, but the Society lacked a full and orderly record of what he (and the ad hoc curators) had done and what conclusions had been drawn. The terse chronological journal records of the weekly meetings, and copies of some of the papers and letters read at the meetings and subsequently copied into the register book, did not constitute the record which Hooke maintained was essential and Council now ordered him to begin to gather together. Hooke's published works do not constitute a coherent corpus, which has had a major influence on those who followed him in science. As Michael Hunter shows below, Hooke usually published a book only when his employment made it necessary or expedient to do so.

On 8 December 1679 a new programme of experiment and publication, and the administrative changes thought necessary to carry it out, were approved by Council. The responsibility for implementing the programme fell on the Curator of Experiments and the Secretaries. Thomas Gale, High Master of St Paul's School, later Dean of York, was the second Secretary. On being asked his opinion of undertaking what was proposed, Hooke answered 'that he would see what he could do in it, but could not as yet undertake it absolutely'.[59] Hooke, for the first time in connection with his duties to the Royal Society, was diffident. Council agreed to pay him £40 at the end of the year (in addition to his £30 salary) as 'an encouragement' to be 'more sedulous' in his experiments and to write treatises on what he had already done.[60] Hooke found it impossible to fulfil all his duties. His weekly demonstrations and experiments continued and he wrote accounts of his philosophical investigations, but he fell short of what was expected of him as Secretary. At the Society's annual elections on 30 November 1682, Hooke lost his place on Council and was replaced as Secretary by Robert Plot, Keeper of the Ashmolean Museum and natural philosopher. Within a week of losing his place on Council and the office of Secretary, Hooke handed over Council's books and the key to the Society's chest to his successor and gave the Society's bonds to the Treasurer, Abraham Hill.[61] It was later discovered that Hooke had not handed in all the documents he had in his possession as Secretary and that his secretarial journal records were incomplete and written in stitched paper books instead of the leather-bound volumes used by Henry Oldenburg, into which Hooke's records had to be transcribed.[62] Hooke's five years as Secretary had not been a success.

In the mid-1680s a new administrative discipline was brought to the affairs of the Society. Hooke continued to be paid his curatorial salary, but other members were appointed ad hoc curators more frequently than before, and in a more orderly manner. On 28 February 1683 the physicians Frederick Slare and Edward Tyson were chosen to make weekly chemical experiments and anatomical dissections respectively, for which they would each receive at the end of the year a piece of silver plate worth £20 as a gratuity.[63] On 6 June 1683 Council decided that Hooke's experimental work should be more closely supervised than before, and that instead of receiving a salary, he should be paid a gratuity every quarter, based on results. At this time Hooke was not keeping a diary, so we do not know what he thought of this decision. After a few months the decision was forgotten by the Society but not by Hooke. In November 1683 Hooke told Council that he intended to write an account of the usefulness and consequences of all the experiments he had performed before the Society and to repeat those that had not been executed or recorded properly.[64] Council was sceptical, demanding a full record of his experiments over the past six months and asking various questions about experiments he had performed over the previous seven years. In the absence of any satisfactory response from Hooke in the following six months, the French natural philosopher Denis Papin carried out demonstrations and experiments for the Society from midsummer 1684 to the end of 1687, for which he was paid regularly at a rate of £30 per annum until the end of 1685 at least.[65] Papin was not appointed to the office of Curator, nor was Hooke formally removed from it, but Papin was in effect a salaried curator during this period.

Although he received no salary during the time Papin was being paid for curatorial duties, on 16 June 1686 Council ordered 'That Mr. Hooke be allowed his arrears for the years 1684 and 1685; and that the treasurer pay him sixty pounds in full till Lady-day last.'[66] Hooke eventually received the arrears on 3 December 1687.[67] The Society continued to pay his salary for only two more years, until Lady Day 1688.[68] This final payment brought Hooke's total salary payments from the Society to £751 13s. 4d.[69] The office of Curator of Experiments effectively disappeared in 1688,[70] but he continued, without a salary, to play an important part in Society meetings, receiving only various expenses and fees for use of his rooms in Gresham College for Council meetings.[71] The last recorded payment to Hooke by the Royal Society is of £6 on 4 December 1702 for six months' use of his room.[72] The accounts are signed by Isaac Newton, who became President of the Royal Society in 1703 soon after Hooke's death.

The sum awarded in 1696 by the Court of Chancery to Hooke as a result of his action against Sir John Cutler was calculated, but the report of it has not survived.[73] In 1695 Hooke was claiming £550 arrears of salary, having received payments amounting to £925,[74] so it is safe to assume that the award in 1696 would have brought his total earnings as Cutlerian Lecturer to about £1,500. On hearing that the Court of Chancery had awarded him the arrears of his Cutlerian salary, Hooke wrote on 18 July 1696, 'I was Born on this Day of July 1635. and God has given me a new

Birth, may I never forget his Mercies to me; whilst he gives me Breath may I praise him.'[75] The fulsomeness of this statement implies that the court decision meant more to Hooke than payment of a sum of money. For more than two decades he had stood up for himself in a struggle against what he saw as unfair treatment by Cutler and the Society in connection with the Cutlerian Lectureship.[76] Nevertheless, throughout this period he had loyally served the Society by his brilliant inventions, imaginative demonstrations, and hard physical work making and using instruments. He may well have seen the court order as a gift from God which was not only a victory in his battle against injustice, but also a vindication of his life as a natural philosopher.

Professor of Geometry

Gresham College was established by the will of Sir Thomas Gresham, who died in 1579. He made his fortune through commerce and trade. (His statement that 'bad money tends to drive out good' has become known as 'Gresham's Law'.) As Queen's Agent he contributed significantly to the Crown's wealth by managing the royal debt and reforming customs practices. He established the Royal Exchange, based on the Bourse in Antwerp, where he lived for many years. According to one of the provisions in his will, his large house lying between Bishopsgate Street and Old Street in the north-east quarter of the City of London (Fig. 6) was to be used as a college to house seven professors, who would each be paid £50 annually for life (provided they remained unmarried) to give free public lectures under the headings of divinity, astronomy, geometry, music, law, physic, and rhetoric. The stipend was generous enough to attract the best from the Universities of Oxford and Cambridge. Gresham's intention was not to establish a similar institution in London, but to make freely available to all London's citizens an education that would lead to the application of knowledge for the general good in areas such as navigation, trade, commerce, medicine, and manufacturing.

In accordance with Gresham's will, funds for the maintenance of the College came from rents paid by tenants at the Royal Exchange. Management of College affairs was shared between the Mercers' Company and the City of London. A Joint Grand Gresham Committee[77] was formed from the two organizations to be trustees of Gresham's will. They created two sub-committees, the 'City Side'[78] and the 'Mercers' Side', each having autonomy in appointing professors: the former was responsible for appointments in the first four disciplines named above, including

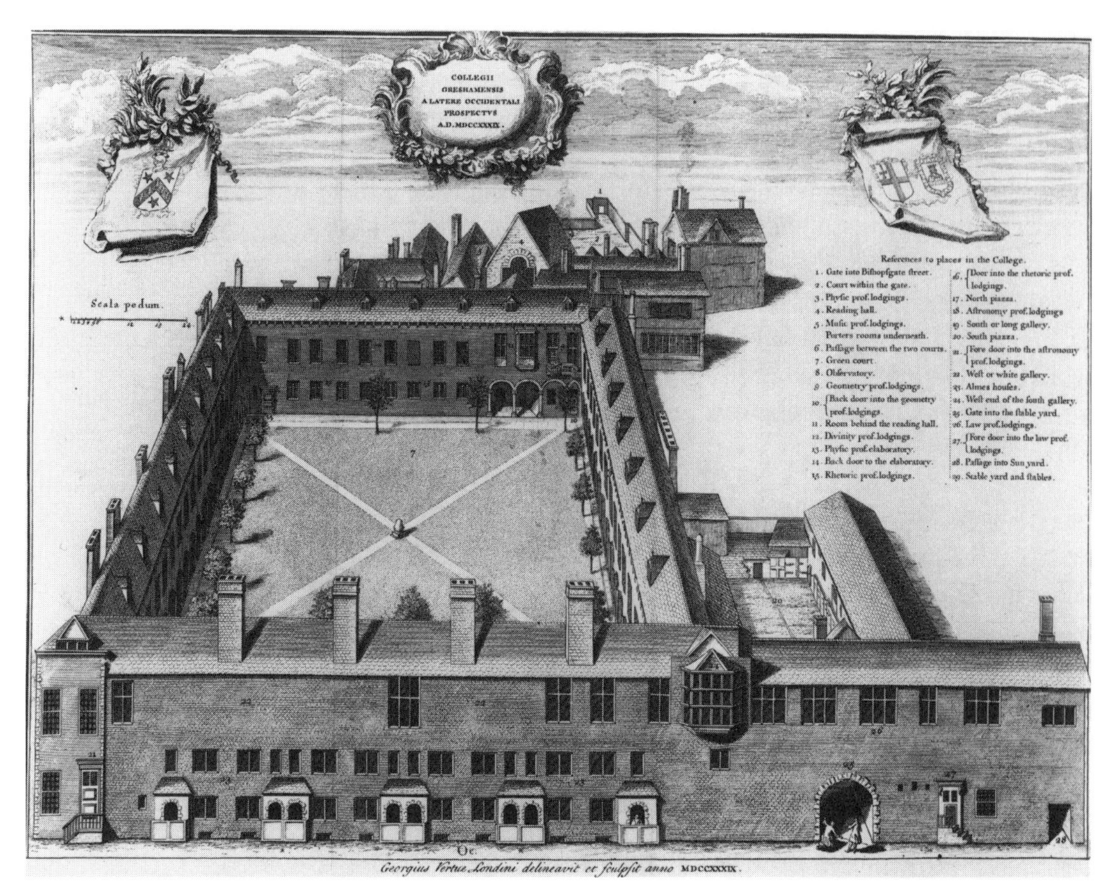

Fig. 6 George Vertue's 1739 engraving of Gresham College. Vertue has the sun shining from the north-west, which is unrealistic. (Guildhall Library.)

geometry. Inaugural lectures by Gresham Professors took place in 1597 and are still given today, in Barnard's Inn Hall, Holborn. The Gresham Trustees are still responsible for fulfilling Sir Thomas Gresham's original intentions, but in a present-day context.

Hooke's appointment as Gresham Professor of Geometry on 20 March 1665 was not straightforward. Some aspects of the means by which he was appointed are still obscure. Laurence Rooke, appointed Gresham Professor of Astronomy in 1652, changed his appointment to Professor of Geometry in 1657, probably because the Geometry Professor's rooms were more suitable for the informal meetings of the men who later formed the Royal Society. Christopher Wren succeeded Laurence Rooke as Gresham Professor of Astronomy. A few days after Rooke's death on 26/27 June 1662 the Gresham sub-committee met to decide which of two candidates, the physician Arthur Dacres and the mathematician Isaac Barrow, should replace Rooke. Gresham appointments were not taken lightly. Candidates were expected to present written testimonies and petitions to the relevant Gresham sub-committee, which usually also sought opinions from members of the Royal Society and from past and present professors at the College and at Oxford and Cambridge Universities. Barrow's and Dacres's petitions and testimonies were read. The sub-committee decided to appoint one of them, but deferred the decision in order to determine 'which was most learned in Geometry'.[79] The issue was resolved by Dacres's withdrawal of his application at the suggestion of Sir Richard Browne.[80] So, on 16 July 1662, Isaac Barrow was appointed Gresham Professor of Geometry, only to announce his resignation the following year to take up an appointment in Cambridge as the first Lucasian Professor of Mathematics in 1664.[81]

In December 1663, around the time that Isaac Barrow decided to leave Gresham College for Cambridge, the Royal Society was trying to get Hooke to take lodgings four days a week in Gresham College so that he would be conveniently placed to prepare experiments to show to the King. When the City Side met on 20 May 1664 to appoint a Professor of Geometry, Arthur Dacres was again a candidate, and so was Hooke, supported by the Royal Society.[82] The report of the meeting of the City Side gives no hint of the deception underlying the appointment of Arthur Dacres:

> two learned persons viz. Dr Arthur Dacres and Mr Robert Hooke being suited for the same, their petitions being Read theire ample Certificates considered and the matter debated The Court proceeded to election and made theyreof the said Dr Dacres to supply the said place of Geomitry Reader in the College.[83]

At the Royal Society Council meeting on 8 June 1664 the validity of Dacres's appointment was questioned 'upon information given, that the lord mayor of London was not of the committee, and yet by his presence had carried the election by a casting vote'.[84] The Society appointed John Wilkins, the lawyer Dudley Palmer, and Daniel Colwall, who was the Society's longest-serving Treasurer in the seventeenth century, to consult with Andrew Ellis, another member whom the Society often asked for advice on legal matters, to investigate the legality of the Lord Mayor's presence in the City Side, and to seek justice for Hooke who, Council was told, had received five votes against Dacres's four. The information given to the Society by the anonymous informer was pertinent, but not completely accurate. A new City Side under the chairmanship of Sir John Lawrence, the Lord Mayor, was appointed by the City's Court of Common Council to consider Hooke's appeal against the earlier decision. The new committee met on 20 March 1665 at Gresham College and found that the nine members legally appointed to the earlier City Side had voted five to four in favour of Hooke. Samuel Foote, Sir Richard Browne, Alderman Thomas Bateman, and Sir William Bateman (two brothers of the Lord Mayor, Sir Anthony Bateman) had voted for Dacres; Sir Thomas Adams, Colonel Neville, Deputy William Flewellen, Deputy John Tivill, and Nicholas Penning had voted for Hooke.[85] Of these, William Bateman, Foote, Adams, Nevill, Flewellen, and Tivill were also at the second meeting of the City Side. Their report of the earlier meeting continued:

> But the then Lord Mayor being present in the said Comittee of 20th May aforesaid, and giveing the vote though not appointed one of that Comittee by the Said Act of Comon Councell was then pleased to Declared [sic] the Election for Dr Dacres. Which mistake occasioning the petition before recited to bee presented to the Comittee now Sitting They after due Consideration of the Principles did unanimously Declare that Robert Hooke was the person legally elected and accordingly ought to enjoy the same with the Lodgings proffits and all accomodations to the place of Geomitry Reader appertaining.[86]

A 'mistake' had been rectified, but was there a cause? Sir John Cutler probably had something to do with it. At the time of Dacres's illegal appointment Cutler had Hooke in mind as his pseudo-Gresham Lecturer on History of Trades, so he had an interest in preventing Hooke's appointment as Gresham Professor of Geometry. Much later, during his litigation against Cutler, Hooke wrote in evidence that in May 1664, soon after his failure to be appointed Gresham Professor and only three weeks before Cutler's proposal first came to the attention of the Royal Society, he met Cutler by accident in a public house. In conversation Hooke mentioned to Cutler his recent disappointment at the hands of the City Side, one of whom was related to Cutler. In reply, Cutler said that Hooke need not be too upset about it because he (Cutler) intended to pay Hooke an equally good allowance to enable him to continue the successful work that Cutler had heard about.[87] It is probable that the man Hooke was referring to was Samuel Foote (or Foot). In 1642 Cutler had married Elizabeth Foot, daughter of Sir Thomas Foot, to whom Samuel was probably related.[88] A conspiracy between Cutler, Foote, and the Bateman brothers to save Hooke for Cutler's sponsorship is the most likely explanation of the 'mistake'. In any case, the episode shows that the Royal Society was powerful enough to overcome an attempt by some leading figures in the City of London to determine Hooke's employment.

Sir Thomas Gresham's will[89] did not specify the duties of the Gresham Professors in detail, so from time to time during the past 400 years successive Gresham Trustees have interpreted Gresham's wishes in different ways.[90] At the time of Hooke's appointment the duties of the seven professors were to read, in the College Reading Hall, one public lecture each week during the four law terms. Each lecture had to be given once in Latin and once in English. In return for his weekly lecture a professor was paid £50 annually, had private lodgings in the College, and the use, in common with the other six professors, of appurtenances such as gardens, courtyards, and stables. Only a celibate man could be appointed and he was required to remain celibate or resign his position. Hooke occupied the Geometry Professor's lodgings in the south-east corner of the quadrangle until his death in 1703. His lectures were to be given each Thursday, in Latin from 8 a.m. to 9 a.m. and repeated in English from 2 p.m. to 3 p.m. The subjects of the geometry lectures were specified as theoretical geometry in Michaelmas and Hilary terms, practical geometry in the Easter term, and arithmetic in the Trinity one.[91]

Having made appointments with due care and diligence (Hooke's failed application was an exception, soon rectified) the College administrators seem to have expected the professors to be similarly diligent in the performance of their not very onerous duties. Although some were, others shamelessly exploited their positions. At a meeting of the Mercers' Side on 15 November 1673, information about various abuses by the professors of their appointments was presented. It was alleged that families were residing in the College and 'unfit meetings' were taking place, which disturbed the rest of the residents. The Mercers' Side decided to ask the joint committee to investigate.[92] Ian Adamson has described what they found and reported in March 1676.[93] Only two professors (one was John Mapletoft, Professor of Physic, the other was Hooke) were resident in the College; the rest, having let their lodgings, were either living at ease in the country or overseas, or pursuing their careers in other places. Thomas Baines, Professor of Music, had let his lodgings and stable to Elias Harvey. Sir Andrew King rented the stable of Walter Pope, the Professor of Astronomy, but lodged in the College's public rooms.[94] Walter Pope's lodgings were either empty or let to Mr Barfoot (it could not be ascertained which). A Mr Crispe, who rented the lodgings of Roger Meredith, Professor of Law, had converted the stable and hay room into a hall and kitchen and made a door and steps out into Broad Street.[95] Hooke's misbehaviour was relatively minor; he had let his stable to a Mr Sutton, who rented the lodgings of Henry Jenks, Professor of Rhetoric, who had let his stable to Dr Croone. Ten years after the 1676 report, little had changed. After several visits by the Trustees to the College they reported that they had 'found the same in great disorder'.[96] On the whole, the professors behaved rather badly and the Gresham Trustees seem to have looked on with weary resignation, but resorting to the law could lead to great expense of time and money. The Trustees had experienced the case of Thomas Horton, Gresham Professor of Divinity, who married in 1651. He refused to relinquish his post and obtained dispensation from a Parliamentary committee to retain it. The dispensation was renewed under Cromwell and under Charles II until it was finally revoked in 1661.[97]

There is clear evidence that by the time of his appointment as Gresham Professor of Geometry in March 1665 Hooke had already given some Gresham Lectures. In a letter to Boyle dated 6 October 1664, Hooke wrote that he is 'this term . . . being engaged to read for Dr Pope'.[98] Abraham Hill, merchant and treasurer of the Royal Society, received and signed for Walter Pope's Gresham salary for three half-years from Lady Day 1663 to Michaelmas 1664.[99] Isaac Barrow gave Pope's lectures until he (Barrow) left for Cambridge,[100] whereupon Hooke took them over. Abraham Hill might have passed on the half-yearly payments of £25 to Barrow and Hooke directly because the monies did not appear in the Royal Society's accounts at that time. Pope was back in London by 26 May 1665, when he signed for his salary for the half-year to Lady Day 1665.[101] Walter Pope was one absent Gresham Professor whose obligation to give lectures was met through the Royal Society and who apparently did not draw his salary when he was away.

In trying to discover how far Hooke went in meeting his obligations to Gresham College we have evidence from his diaries and Royal Society and Corporation of London records that he hardly spent a night away from Gresham College, except during the plague epidemic of 1665 and a few visits to his family in the Isle of Wight. Indeed, he was rarely more than a mile or two, or an hour or two, out of the College. In contrast, the available evidence about his performance of Gresham Lectures is scanty and complicated. He gave his Cutlerian Lectures in the same place (the public Reading Hall at Gresham College)[102] and on the day of the week when the Royal Society met.[103] The time specified for his Gresham lectures in English (2–3 p.m.) coincides with the time of his Cutlerian Lectures, which preceded the 3 p.m. start of the Thursday meetings of the Royal Society. Although his Cutlerian Lectures were intended to be given only in the vacations, the intention was not always carried out. His programme of Gresham Lectures was determined by law terms, which varied in dates and duration according to the ecclesiastical calendar. Although, as Michael Hunter discusses below, Hooke had good reasons to publish some of his Cutlerian Lectures, his Gresham Lectures remain unpublished, except for those included by Richard Waller in Hooke's *Posthumous Works*.[104] In his accounts of Sir John Cutler's patronage and the Royal Society, Michael Hunter has presented detailed evidence about Hooke's Cutlerian Lectures, which has been used here in an attempt to make a distinction in Hooke's diary entries between

Cutlerian and Gresham Lectures in order to reveal something of his performance in giving his Gresham Lectures.[105]

It soon becomes clear from Hooke's diaries that the most significant aspect in assessing his performance of Gresham Lectures is not whether he was present to give them (he almost always was) but whether anyone was present to hear them. Many of his diary references to what we now conclude were Gresham Lectures refer either to the small number present or to the complete absence of an audience, as Ian Adamson has illustrated.[106] These entries show that sometimes Hooke waited in the hall until 3 p.m. to see if anyone came. At other times his niece Grace, or Harry Hunt, the Royal Society's operator, was on hand to call Hooke when anyone arrived, or to speak to latecomers if he had left the Reading Hall and was at the Royal Society meeting. Hooke would not give a Gresham Lecture after 3 p.m.—the starting time of the Thursday meetings of the Royal Society. Almost all Hooke's diary entries about his Gresham Lectures refer either to those that did not take place because nobody was present to hear them, or to those that did take place but had very few listeners. There are exceptions, usually when he lectured to masters and pupils from Christ's Hospital: 'Noe auditors [morning] . . . between 2 & 3 Read 3/4 hour Lecture, in the hall Paget & 2 other men and 40 children.'[107] Only once in his diaries does Hooke explicitly identify a lecture as a Gresham Lecture: 'Read not for Sir T. Gresham, nor for Royal Society,'[108] and only once does he mention the topic of what can be assumed was a Gresham Lecture: 'Attended morning Lecture, none came, not one . . . Read Lecture of 1st and 2nd propositions of Euclid.'[109] Occasionally he recorded his suspicions that someone in the audience at one of his lectures was sent there either by the Gresham Trustees or by Sir John Cutler to see whether or not he was doing what he was paid to do. His comment 'A fellow with a blew apron layd asleep all the time. there should have been a lecture, I suppose a spy' refers to what was possibly a Cutlerian or afternoon Gresham Lecture which he did not read because there was no other person present.[110]

In contrast to Hooke's close and continuous engagement with the Royal Society, the Gresham Trustees left him alone. His response was to show in his attention to his Gresham obligations considerably more virtue than many of the other Gresham Professors, who used their appointments as sinecures and pursued their interests elsewhere, at home or abroad. Of course, Hooke's other main interests happened to be in or around Gresham College, so it was more than convenient for him to occupy his lodgings, where, although celibate, his private life was rackety at times.[111] Nevertheless, the little evidence about his Gresham Lectures that has been found shows that he was diligent in turning up or being

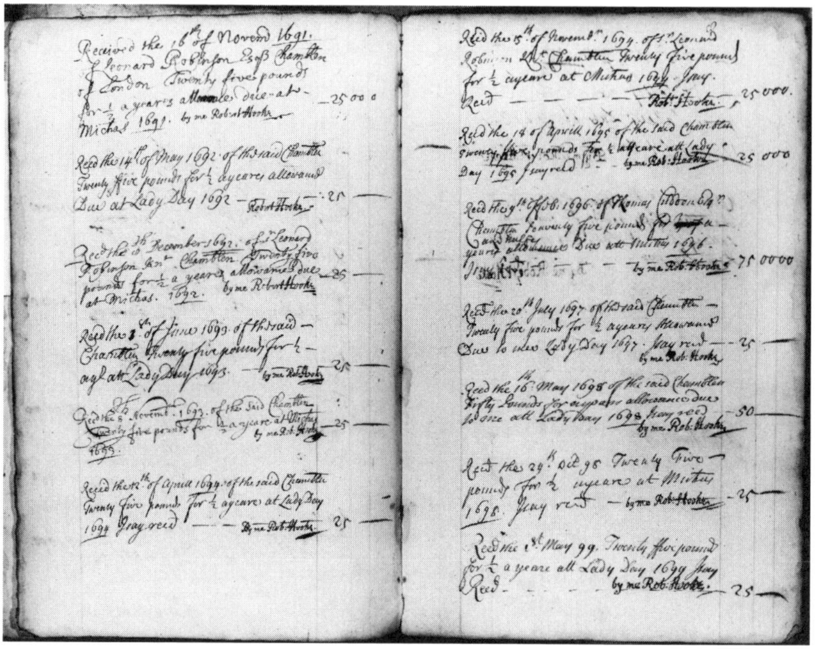

Fig. 7 Robert Hooke's signatures for receipt of payments of salary as Gresham Professor of Geometry. Gresham College Accounts Acquittance Books III (1683–1707) folios 7v and 8r. (Corporation of London Records Office.)

readily available to read them, even though they were often not well attended and sometimes had no audience at all. On their part, the Gresham Trustees paid Hooke's salary regularly, according to their account books. He went regularly to the City Chamberlain's office in Guildhall to collect and sign for his Gresham salary until 23 July 1702, when he collected his year's salary to Lady Day 1701 (Fig. 7).[112] His signature over the later years shows a gradual change, which is evidence of increasing infirmity towards the end of his life. His total earnings as Gresham Professor of Geometry from Lady Day 1666 to Lady Day 1701 amounted to £1,750.

City Surveyor

The Great Fire of London began in a baker's shop in Pudding Lane in the early hours of Sunday, 2 September 1666. At first it was no different from many other fires in the city that had quickly been put out, but this time a strong wind spread the flames rapidly to the north and west across a city that had been made tinder-dry by a hot summer.[113] After four days and nights the Fire died down as quickly as it had begun, leaving most of London a desolate, charred, and smoking ruin. Its former streets, alleyways, and cramped courtyards were buried in rubble. So many buildings had been reduced to the ground that it was possible to see across the city from east to west and the Thames was visible from Cheapside.[114] Within the walls 85 per cent of the city was destroyed, and outside, to the west, the destruction extended beyond the Fleet River as far as Temple Gardens. Domestic, commercial, business, and parish life were ruined. More than 50,000 citizens were homeless. St Paul's Cathedral, 84 of the 109 parish churches, 44 livery company halls, Guildhall, the Royal Exchange, the Custom House, gateways, prisons, law courts, and the tools and wares of craftsmen, shopkeepers, and tradesmen were beyond use. As suspicions and rumours about how the Fire started took hold among the citizens, the authorities had to act quickly to maintain law and order and show that they were doing something towards the recovery of trade and commerce and the rebuilding of the city. The Fire brought bankruptcy and despair to many, but it brought wealth to others, including Hooke. Until recently[115] his work as City Surveyor has been largely ignored and generally misunderstood, despite abundant evidence about it in his manuscripts in the Corporation of London Records Office; a brief description of them was published in 1922.[116]

On Thursday 6 September 1666 when parts of London were still smouldering and likely to catch fire again, the City's Court of Aldermen (Fig. 8) met in Gresham College in the north-east corner of the city which had escaped the worst effects of the Fire, instead of in Guildhall, which was burnt out. The City[117] wasted no time before starting the formidable task of keeping order and gaining the confidence of citizens. Orders were issued that the Lord Mayor and Sheriffs who had lost their houses should make use of rooms in Gresham College.[118] The City was in no doubt that occupation of the College was not only necessary but proper, and the professors were told to move out. Despite severe overcrowding and the urgent need for accommodation, the City decided that Hooke could continue to occupy his rooms and the Royal Society could also continue to use the College for its meetings and Repository.[119] In allowing this continued occupation, the rulers of the City showed that they regarded Hooke and the Royal Society as important to them.[120] Hooke, already well known to the City through his Gresham appointment, his sponsorship by Sir John Cutler, and his connections with the City's guests, the Royal Society, was now living and working among the Ward Aldermen and Deputies who governed London, and their officers. He was well placed to put himself forward as somebody who could be useful to them in their endeavours to rebuild the city. He knew to whom he should speak and what to offer—he had acted similarly ten years earlier at Wadham College, when he eagerly showed his attempts at making flying machines to John Wilkins. He now seized the first opportunity that

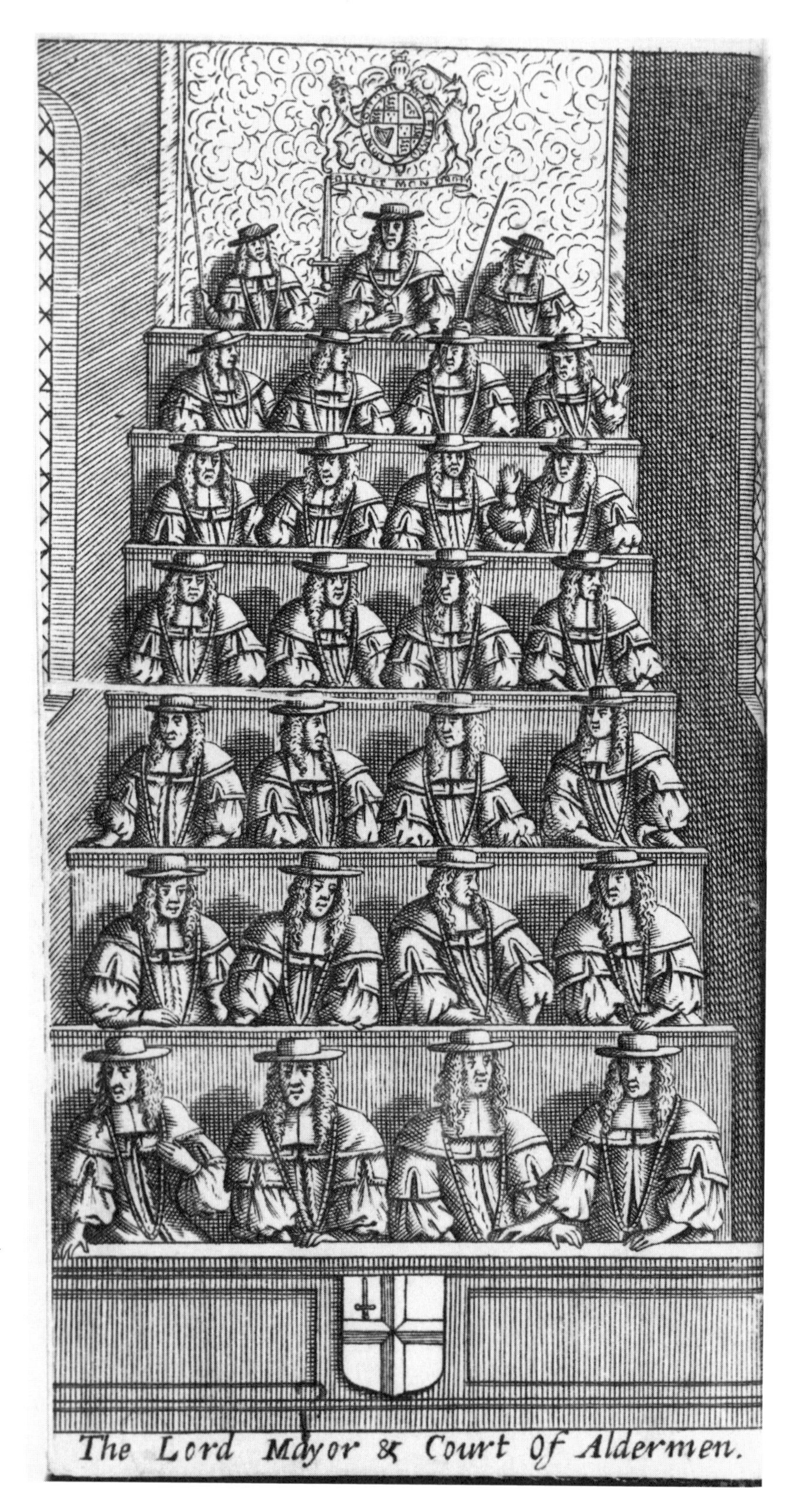

Fig. 8 The Lord Mayor (top centre), Sheriffs (top left and right), and the Court of Aldermen of the City of London, from T. de Laune, *The Present State of London* (1678). (Guildhall Library.)

came his way after the Fire by drafting a plan for rebuilding London and showing it to the City.

Plans for new layouts were quickly prepared by at least five others, some of whom produced more than one.[121] Hooke showed his plan or 'model' first to the City, in particular to Sir John Lawrence, who had been Lord Mayor in 1665 and chairman of the meeting of the City Side which rectified the improper election of Dacres as Geometry Professor. Hooke then showed his plan to the Royal Society at its meeting in Walter Pope's lodgings in Gresham College on 19 September 1666:

> Mr. Hooke shewed his model for rebuilding the city to the society, who were well pleased with it; and Sir John Laurence, late lord mayor of London, having addressed himself to the society, and expressed the present lord mayor's [Sir Thomas Bludworth] and aldermen's approbation of the said model, and their desire, that it might be shewn to the King, they preferring it very much to that, which was drawn up by the surveyor of the city [Peter Mills]; the president answered, that the society would be very glad, if they or any of their members could do any service for the good of the city; and that Mr. Hooke should wait upon them with his model to the King, if they thought fit to present it: which was accepted with expressions of thanks to the society.[122]

Hooke's plan is now lost. Waller records that he heard it was in the form of a rectangular grid of streets, but he is not certain.[123] Reddaway intriguingly, although not unambiguously, records that Hooke's plan has survived, but he does not say where it can be found.[124]

None of the new layout plans for London was implemented. The need to resume normal domestic and commercial life was overwhelming. Neither time nor money was available to buy land for building a grand new city. The City decided that regulated private rebuilding should go ahead, mainly on the old foundations. Expediency, however, was tempered by an intention that the buildings in the new city should be more substantial and its markets, quays, and streets more regular than before. When the City was informed at a Privy Council meeting that the King had appointed Christopher Wren (Fig. 9), Hugh May, and Roger Pratt as his Commissioners for Rebuilding to work with the City's nominees on a survey of the ruins,[125] they quickly nominated Hooke, Peter Mills, and Edward Jerman.[126] Although Hooke was the youngest and least experienced of the City's nominees, he had carried out experiments for the Royal Society on the strengths of different kinds of wood[127] and had investigated Kettering-stone (see Fig. 4) as a building material for Wren.[128] Further evidence of Hooke's interest in and knowledge of building

Fig. 9 Christopher Wren (1632–1723), engraved by Edward Scriven. (Museum of the History of Science, Oxford.)

construction is in a report[129] he presented to the Royal Society on 8 June 1664 about damage done to a building in Piccadilly during a thunderstorm the previous afternoon, in which he shows detailed knowledge of building construction. He visited the site, examined the building, spoke about the storm with brickmakers, carpenters, and others who witnessed it, and based his report on what he saw and what he heard. He would later, in a similar manner but for more than scientific curiosity, stand before many ruined buildings, take evidence from witnesses, and report formally on what he heard and saw. For several years Hooke had worked with and supervised craftsmen making instruments, first for Willis and then for Boyle and the Royal Society. He had gained an understanding of different craft and trade practices and their practitioners that would be of value when he came to supervise the rebuilding of London.

Reddaway has said that the King showed foresight that was later justified in appointing Wren as one of the Commissioners for Rebuilding.[130] In nominating Hooke as one of their Surveyors, the City showed similar foresight. He was already well known to the Lord Mayor and Aldermen of the City, whose mercantile and business instincts probably led them to see him as an able and willing 31-year-old in need of an income and who alone could give the City's rebuilding programme an intellectual presence similar to that of Wren, one of the King's Commissioners. The burgeoning friendship between the two men and their common scientific interests would have been recognized by the City as important for the good working relations with the King that were so necessary for the speedy reconstruction of the city's buildings and the resumption of its trade and commerce. The importance of the Wren/Hooke partnership to the City's rebuilding was also discerned in the Royal Society; the Secretary, Henry Oldenburg, said in a letter to Boyle that 'by the care and management of Dr Wren and M. Hook' the survey and measurement of the foundations would be 'exactly registred' and 'then the method of building will be taken into nearer consideration and within a short time resolved upon'.[131] The long partnership (in science, architecture, and building construction) between Wren and Hooke was very important to both men and is at last receiving proper attention.[132]

The City relied again on Hooke for technical expertise when it nominated him, Peter Mills, and Edward Jerman to work with the King's Commissioners and decide on the building regulations for the new Act of Parliament:

> This Court doth nominate & appoint Mr [blank] Hooke of the Mathematicks in Gresham house Mr Peter Mills & Mr Jermyn from time

> to time to meete & Consult with Mr May Dr Wren & Mr Pratt Commissioners appointed by his Majesty concerneing the manner forme & highth of Buildings in this City the Scantlings of Timber removeing of Conduits and Churches and Alteration of the Streetes And it is ordered that from time to time they report such their Consultation to this Court and give noe Consent or make any Agreement therein without the speciall Order of this Court.[133]

The latter proviso was a formality; the City's three nominees were left to decide matters on its behalf. In effect, this meant Hooke alone. Jerman seems to have played no part. Mills was growing old and would soon become ill.[134]

The directions given to Hooke by the City were not yet accompanied by an official appointment. No mention of a salary or other form of payment to him has been found in the City records up to this time. Nevertheless, Hooke was active during the winter months of 1666/7 on work the City had nominated him to undertake. In the four weeks beginning on 4 October 1666 Hooke received orders to oversee the compilation of what would now be called a land information system for London, to draw up building regulations for an Act of Parliament governing the rebuilding of the city, and (from the Royal Society) to conduct experiments to find which material would make the best bricks. The urgency and magnitude of these tasks were too much even for Hooke. Only the second was completed. The first has only recently been seriously attempted and work on the third is still in progress. Despite knowing about his unofficial work for the City, the Gresham Trustees at this time did not hesitate to give him even more work to do that lay outside his professorial duties.

At their meeting on Friday, 2 November 1666, the Gresham Trustees decided to get an estimate of the cost of rebuilding the Royal Exchange. They sent Hooke to visit the site and make a report. Again Hooke eagerly made the most of an opportunity. Two weeks later he submitted his report, but the Gresham Trustees did not think much of his economical proposal to use as much old material as possible in rebuilding the Exchange. As Ann Saunders has said,[135] they wanted to demonstrate to the world the City's ability to overcome a disastrous fire without ceasing trading, and to take pride in a glorious new Royal Exchange. By that decision they set in motion a decline in the fortunes of the Mercers' Company, which led to near-bankruptcy around 1700 (the Mercers suffered more than the City because the City's regular income was supplemented by the Coal Tax). As we have seen,[136] at the end of the seventeenth century the Gresham Trustees were seeking approval from Parliament to demolish

Gresham College and Hooke was supporting the Royal Society in its attempts to retain their accommodation there. He no doubt remembered that the advice he had given to the Gresham Trustees more than thirty years earlier had been spurned. His estimate then of the cost of rebuilding was in the end exceeded more than tenfold. Such extravagance by Gresham's Trustees, contrary to Hooke's advice, had brought about the position they now found themselves in, and Hooke was indignant that the Royal Society should be inconvenienced in consequence.

Throughout the winter months of 1666–7 meetings took place between the City's three nominees, the King's Commissioners, the City, and the Privy Council. On 8 February 1667 the Royal Assent was given to the first Act of Parliament for rebuilding.[137] This Act incorporated the building regulations devised by Hooke, Mills, Jerman, and the King's Commissioners, but it left the City to determine the widths of the streets to be enlarged under the Act, subject to the King's approval. Hooke and Mills were ordered[138] to assist, and by 13 March 1667 the City was ready to present to the King its Acts of Common Council governing the street widening:

> His Majesty haveing heard the two Acts of the Comon Councell read distinctly to him, of the 26th and 27th of ffebruary last, the Map of the Citty lying before him, his Majesty lookeing upon the lines drawne out in the said Map according to the Orders mentioned & deliberating & discoursing much thereupon; his Majesty doth fully approve & commend all the Particulars mentioned in the said Orders with these Animadversions upon some of them.[139]

The King appointed his Commissioners to be ready at all times to assist the City and its three nominees. He made seven animadversions on the City's proposals. They were not all feasible under the Rebuilding Act, but the City authorities did what they could to meet them. Those that were feasible were incorporated in the City's Act of Common Council of 29 April 1667.[140]

The City's confidence and trust in Hooke during the hectic months from September 1666 are remarkable, especially as he had held no City office until, at a meeting of the Court of Aldermen in March 1667, he and Peter Mills were sworn as the City's Surveyors of New Buildings. Jerman was absent, John Oliver,[141] whom the City had also intended to be appointed, asked to be excused. Six days later the City sealed an Instrument authorizing Mills and Hooke to stake out the streets.[142] Hooke was about to embark on another onerous employment that would test him to the utmost. He showed his new double reflection telescope to the Royal Society at its meeting the day following his order to stake out the streets of London, but reported that the air had been for a good while so thick that he had not been able to observe the stars.[143] Here is an example of his incessant interest in and observation of all things around him, whether down amid the rubble on the streets or above in the heavens. He was now about to be engaged in the personal lives of thousands of individual citizens in four main tasks as City Surveyor: to stake out new, straightened, and widened streets and the foundations of private buildings; to certify areas of ground taken by the City for streets, new quays, markets, etc.; to investigate disputes between citizens arising during rebuilding and recommend how they should be settled; and, in connection with the City's own rebuilding programme, to report on contractors' estimates, bills of quantities, and the quality of workmanship and materials. What he did in each of these activities will now be briefly described.

On Monday, 27 March, only four days after the City authorized the staking out of streets, Hooke and Peter Mills began by realigning and widening Fleet Street.[144] An account written by the City's Clerk of Works shows that, in the first week, the workmen assisting Mills and Hooke were paid for each of the seven days, and that six carpenters and seven labourers used 1220 feet of timber for stakes, which a carter carried around the City for the Surveyors.[145] The progress of staking out streets can be followed in the accounts of the Clerk of Works. In only nine weeks most of the streets had been staked out,[146] but work continued intermittently for the next few years in response to specific requests to the Court of Aldermen from individuals, groups of neighbours, and institutions.

The procedure to be followed by anyone rebuilding a particular private property (the person responsible for the rebuilding might be, for example, a lessee, tenant, or landowner) was defined in the Act of Common Council dated 29 April 1667.[147] The sum of 6s. 8d. for each foundation was paid into the Chamber[148] where the name of the applicant and the number and locations of the foundations were entered into Chamberlain's Day Books (Fig. 10). The applicant was given a receipt which he or she took to one of the Surveyors and made an appointment for the foundation to be staked out, measured, and surveyed. The applicant was responsible for clearing all rubbish from the site beforehand. The Surveyor identified the old foundations by inspection and, if necessary, by listening to evidence presented by his client, neighbours, and any other witnesses he might choose to call on. Deeds giving dimensions were sometimes produced and examined by the Surveyor on site as additional evidence. When satisfied that the old foundations had been located on the ground, the Surveyor staked out the building lines

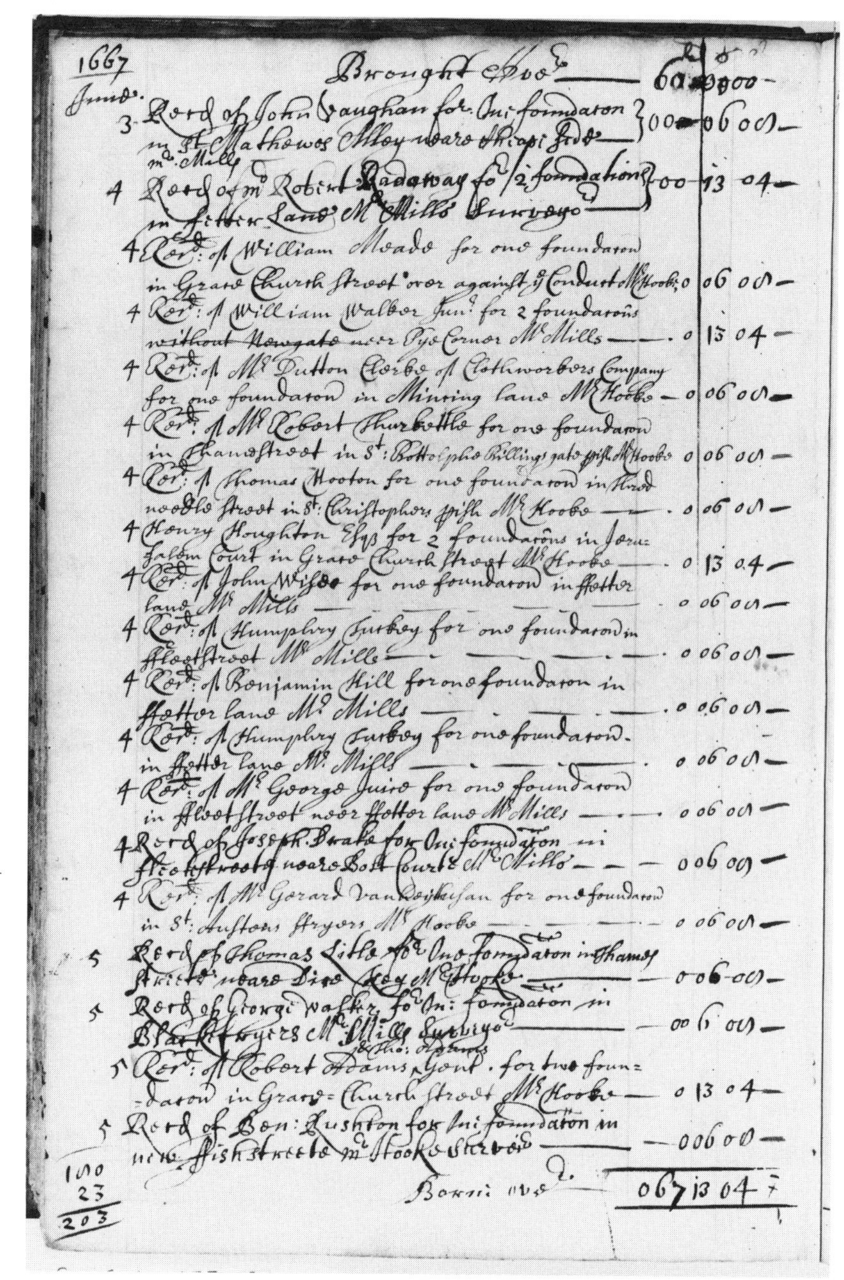

1667 Brought over — 60 00 00
June 3 Recd of John Vaughan for one foundation in St Mathewes Alley neare [illegible] Mr Mills — 00 06 08
4 Recd of Mr Robert Radaway for 2 foundations in Fetter Lane Mr Mills Surveyor — 00 13 04
4 Recd of William Meade for one foundation in Grace Church street over against the Conduit Mr Hooke — 0 06 08
4 Recd of William Walker Junr for 2 foundations without Newgate neer Pye Corner Mr Mills — 0 13 04
4 Recd of Mr Dutton Clerke of Clothworkers Company for one foundation in Mincing lane Mr Hooke — 0 06 08
4 Recd of Mr Robert Thurkettle for one foundation in [illegible] in St Botolphs Billingsgate parish Mr Hooke — 0 06 08
4 Recd of Thomas Hooton for one foundation in Threadneedle street in St Christophers parish Mr Hooke — 0 06 08
4 Henry Houghton Esqr for 2 foundations in Jerusalem Court in Grace Church street Mr Hooke — 0 13 04
4 Recd of John Wilde for one foundation in ffetter lane Mr Mills — 0 06 08
4 Recd of Humphry Turkey for one foundation in ffleetstreet Mr Mills — 0 06 08
4 Recd of Benjamin Hill for one foundation in ffetter lane Mr Mills — 0 06 08
4 Recd of Humphry Turkey for one foundation in ffetter lane Mr Mills — 0 06 08
4 Recd of Mr George Juice for one foundation in ffleetstreet neer ffetter lane Mr Mills — 0 06 08
4 Recd of Joseph Drake for one foundation in ffleetstreete neare Bolt Court Mr Mills — 0 06 08
4 Recd of Mr Gerard Vandeyckhan? for one foundation in St Austins ffryers Mr Hooke — 0 06 08
5 Recd of Thomas Little for one foundation in Thames Streete neare Dice Key Mr Hooke — 0 06 08
5 Recd of George Walker for one foundation in Blackfryers Mr Mills Surveyor — 00 06 08
5 Recd of Robert Adams Gent & Tho: [illegible] for two foundations in Grace Church street Mr Hooke — 0 13 04
5 Recd of Ben: Rushton for one foundation in new ffishstreete Mr Hooke Surveyor — 00 06 08
Borne over — 067 13 04
100
23
203

Fig. 10 The record of payments made by citizens to the City Chamberlain in order to have their foundations staked out, measured, and certified by the City Surveyors. The City charged at a rate of 6s. 8d. for each foundation. Ex-Guildhall Library Manuscript 275, folio 4v. (Corporation of London Records Office.)

and party walls, allowing for any street widening, and measured the lengths of the boundary lines. He recorded in his survey book such information as the date, the client's name, the location and dimensions of the site, the number of foundations, the names of neighbours, and usually a sketch. This procedure was a very common public performance in the City. The certificate was probably written soon afterwards from information recorded in the survey book, and handed to the applicant. Only then could rebuilding begin.

Owners were often desperately anxious to start rebuilding. Peter Mills and Hooke could not keep pace with the rate at which payments were coming into the Chamber, so delays of one or two months built up. The backlog became worse at the end of July 1667, when Mills fell ill for a few weeks. John Oliver then stood in for Mills as he said he would,[149] but either the City or Oliver delayed an official appointment until 28 January 1668, when he was sworn as the third City Surveyor.[150] Mills, Hooke, and Oliver then shared the work which the City had intended to be undertaken by four Surveyors, but Mills was ailing. He carried out his last survey on 19 July 1670[151] and died within the next three months.[152] He was not replaced. Thereafter Hooke and Oliver together shared the duties of City Surveyor.

Very few of the certificates issued to individuals can now be located.[153] Each Surveyor recorded details of his surveys in his own book. Taken together they contained details of more than 8,000 foundations, aides-mémoires, and records of other official activities, but all of them are now lost. Mills and Oliver each handed in to the City ten survey books, which were transcribed into four volumes in the eighteenth century.[154] Facsimiles of the transcriptions have been published by the London Topographical Society.[155] Hooke did not hand in his survey books, despite orders to do so,[156] so they were not transcribed and are also now lost. It is probable that Hooke refused to hand them in to the City because they were useful to him, not only in his surveying, but also for other notes he made in them as he went daily about the streets, taverns, and coffee shops of London.[157] Their absence from the archives goes some way towards the almost complete neglect until quite recently of Hooke's surveying compared with the attention given to the work of Peter Mills and John Oliver in the city, and to Hooke's other employment by Sir John Cutler, Gresham College, and the Royal Society.

It has been possible to make a reasoned estimate of the number of foundations set out and certified by Hooke, even though none of his certificates or survey books is available. His diary gives very little indication of the magnitude of this work because entries begin in 1672, by which time more than 95 per cent of the foundation surveys had been completed. Samples of evidence from the Day Books and elsewhere have been used to show that Hooke staked out almost 3,000 foundations. In particular, during the months of March, April, and May 1669 he staked out and certified at least 90 foundations each month; working six days a week, he would have spent at least three hours each day on foundation surveys alone.[158]

Hooke's second task as City Surveyor was to certify areas of ground lost by private citizens. Compensation payments were made for private land taken by the City for new and widened streets, new markets, wharves alongside the new Fleet Channel, and quays and wharves along the north bank of the Thames. The amount of compensation paid by the City depended on the location of the site, the area of ground taken away, and any melioration of loss brought about by rebuilding. The usual rate of compensation payment was five shillings per square foot. The Surveyors were responsible for measuring the dimensions of each parcel of lost ground and calculating and certifying its area. The certificate (Fig. 11)[159] was handed to the citizen, who took it to the City Chamberlain's office, where warrants for payment of compensation were issued. The Rebuilding Acts[160] allowed compensation for private land taken by the City to be paid from monies raised by a tax on coal (often referred to as 'Coal Monies' or 'Coal Tax' in City documents) levied specifically to pay for the rebuilding of London. The City was also empowered to call on counsel and other assistance, and to pay for such services from the Foundation Cash. The City Surveyors were ordered to attend meetings of the City Lands Committee, which was responsible for administering the compensation payments.[161] More than 650 area certificates written by Peter Mills, Hooke, and John Oliver are extant in the City's archives.[162] From these and other evidence it has been possible to estimate that, from the beginning of 1668 to the end of 1687, Hooke wrote about 350 area certificates of which the highest annual total was 94 in 1671.[163]

Hooke's third task was to recommend how building disputes between neighbours and allegations concerning infringement of the building regulations should be settled. Before the Fire, disputes between neighbours and allegations of irregular building had been investigated and reported in writing (that is, viewed) by the City Viewers (appointed from among the master craftsmen) and the Alderman and Deputy of the Ward where the site in dispute was located.[164] The new legislation and the intensity of rebuilding required viewers with more knowledge and different abilities from those traditionally found among the City Viewers, so the City Surveyors were given the responsibility of viewing new buildings.[165] When a neighbour began building a party wall according to the Acts, an intermixture of interests often arose. A neighbour who formerly had a room at an upper level which lay in part (or sometimes wholly) over a neighbour's room at the next lower level would lose that interest when the vertical party wall was built according to the Rebuilding Acts. Multiple intermixtures, in both directions and at different levels (sometimes below ground level in basements and cellars)

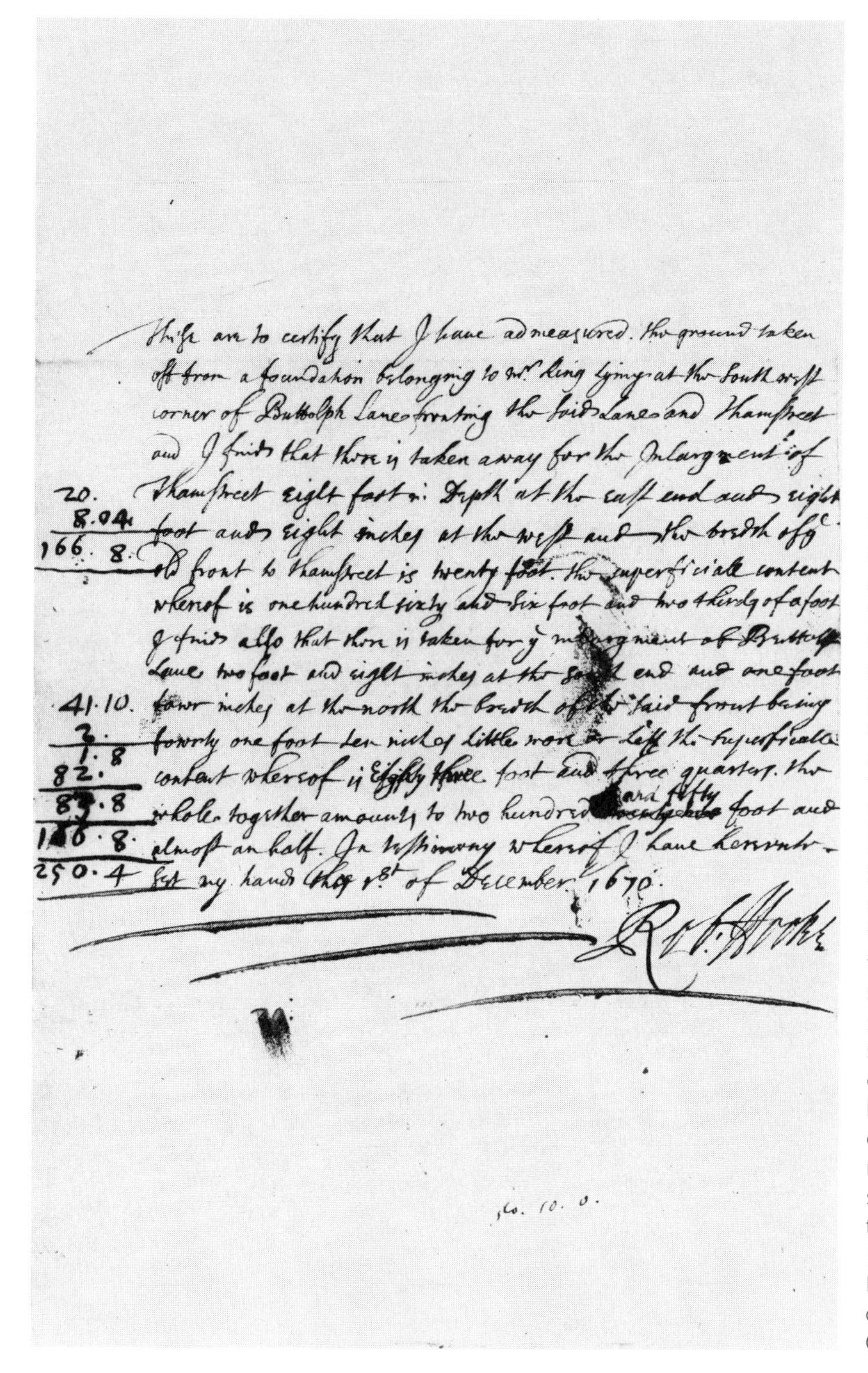

These are to certify that I have admeasured the ground taken
off from a foundation belonging to Mr. King lying at the South west
corner of Buttolph Lane fronting the said Lane and Thamestreet
and I find that there is taken away for the Inlargement of
Thamestreet eight foot in Depth at the east end and eight
foot and eight inches at the west and the breadth of the
old front to Thamestreet is twenty foot. the superficiall content
whereof is one hundred sixty and six foot and two thirds of a foot
I find also that there is taken for ye inlargement of Buttolph
Lane two foot and eight inches at the south end and one foot
four inches at the north the breadth of the said front being
fourty one foot ten inches little more or less the superficiall
content whereof is Eighty three foot and three quarters. the
whole together amounts to two hundred and fifty foot and
almost an half. In testimony whereof I have hereunto
set my hand this 1st of December 1670.

Rob: Hooke

20.
8.04
166. 8
41.10.
2.
1.8
82. 8
83.8
166.8
250.4

50. 10. 0.

Fig. 11 Hooke's certificate for two areas of ground taken away from a foundation lying at the south-west corner of Botolph Lane at its junction with Thames Street for the widening of Botolph Lane and Thames Street. In the left-hand margin is Hooke's calculation of the two areas and their sum, using duodecimal fractions. The total area taken away is 250⁴/₁₂ square feet. A clerk's note at the bottom shows that the amount of compensation paid by the City was £50 10s. 0d., corresponding to a rate of 4s. 0d. per square foot, rounded up to the nearest 10s. (Comptroller's Deeds Box K/K/6, Corporation of London Records Office.)

could occur. The Surveyors[166] had to determine the net gain or loss by each neighbour and its value. Settlement was customarily by deciding the moieties of the cost of building the party wall (normally shared equally) or by a cash payment. Nearly always Hooke was able to say in his reports that the parties had agreed to his settlement proposals

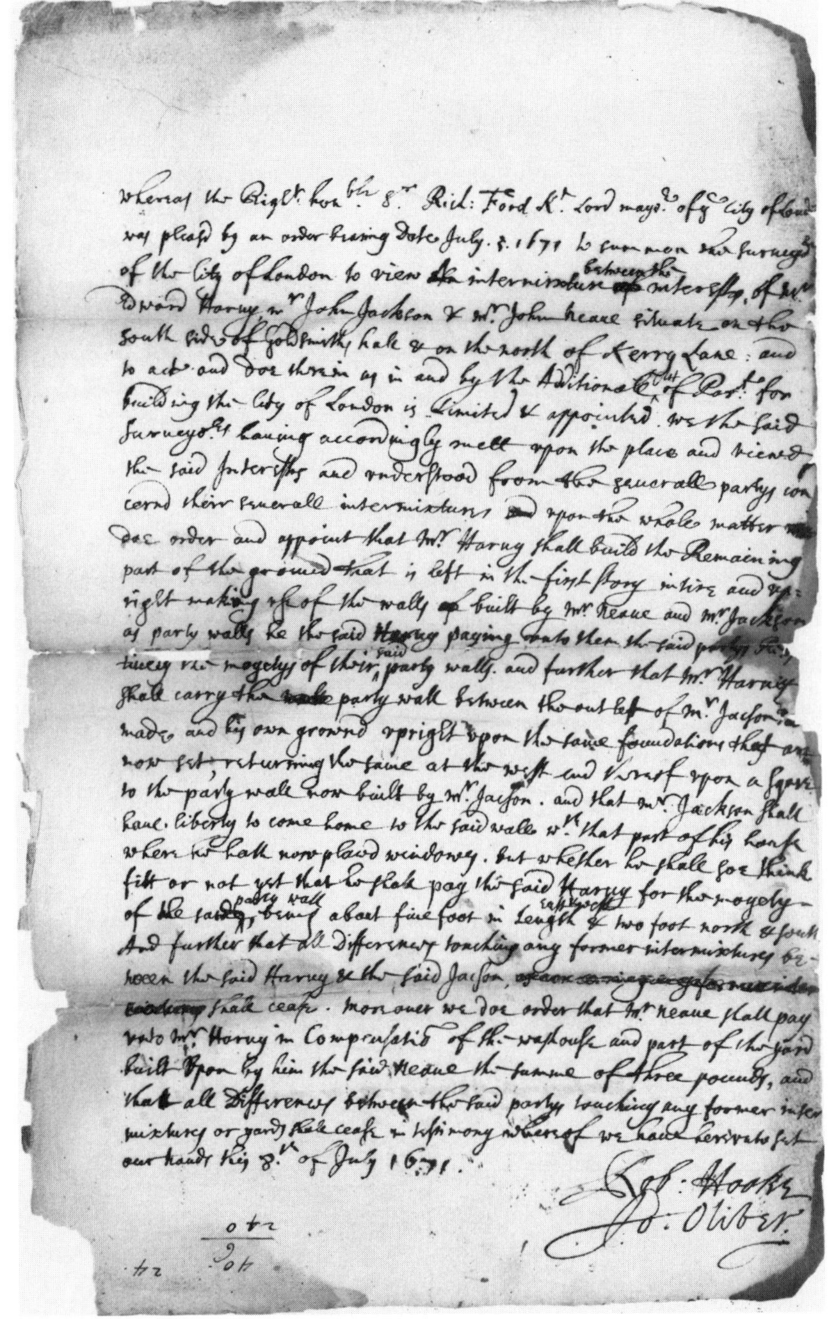

Whereas the Right hon.ble Sr Rich: Ford Kt. Lord mayor of ye City of London was pleased by an order bearing date July. 5. 1671 to summon the Surveyors of the City of London to view the intermixture between the interests of Mr Edward Harvy, Mr John Jackson & Mr John Neave situate on the South side of Goldsmiths hall & on the north of Kerry Lane: and to act and doe therein as is and by the Additionall Act of Parlt. for building the City of London is limited & appointed. we the said Surveyors having accordingly mett upon the place and viewed the said Interests and understood from the severall partys concerned their severall intermixtures and upon the whole matter doe order and appoint that Mr Harvy shall build the Remaining part of the ground that is left in the first story intire and upright making use of the walls built by Mr Neave and Mr Jackson as party walls he the said Harvy paying unto them the said partys severally the moyety of their said party walls. and further that Mr Harvy shall carry the party wall between the outlet of Mr Jackson made and his own ground upright upon the same foundations that are now sett, returning the same at the west end thereof upon a square to the party wall now built by Mr Jackson. and that Mr Jackson shall have liberty to come home to the said walls wth that part of his house where he hath now placed windowes. but whether he shall soe think fitt or not yet that he shall pay the said Harvy for the moyety of the said party wall being about fine foot in Length & two foot north & South And further that all differences touching any former intermixtures between the said Harvy & the said Jackson, [illegible] shall cease. Moreover we doe order that Mr Neave shall pay unto Mr Harvy in Compensation of the washouse and part of the yard built upon by him the said Neave the summe of three pounds, and that all Differences between the said partys touching any former intermixture or yards shall cease. in testimony whereof we have hereunto sett our hands this 8th of July 1671.

Rob: Hooke
Jo: Oliver

Fig. 12 A report of a view by Hooke and Oliver of a dispute about intermixture of interests. (Miscellaneous MS 92.73, Corporation of London Records Office.)

(Fig. 12). Disputes over party walls were not confined to the intermixture of interests. The first to rebuild might infringe the Acts by not leaving holes in the brickwork on the neighbour's side of a party wall to support floor joists, or perhaps by building the wall out of plumb, or of insufficient thickness in relation to the height of the building.

Disputes over rights of light and of access, routes of drainage gullies, fire hazards, and encroachments (even below ground level where one neighbour would attempt to excavate below a party wall to acquire cellar space beneath the adjacent property) were settled by the Surveyors through daily visits to the sites and discussions with disputants and other witnesses. Complaints made by citizens to the City's Court of Aldermen about irregular building, buildings for businesses or trades inappropriate in the particular location, shop fronts projecting into the street, the use of fir instead of deal for load-bearing woodwork, and the use of anything other than brick or stone for the main load-bearing walls had to be viewed and settled quickly.

Views were more complex and time-consuming than certifying foundation surveys and areas of lost ground. Conflicting evidence of how things were before the Fire sometimes had to be reconciled by the Surveyor. Hooke often took oral evidence from neighbours at the site about former passageways, windows, and gullies in the presence of the disputants, who would bring supporters and documentary evidence of their own to make their case. Sometimes he would require sworn evidence, but this could delay settlement by days or weeks, so he generally avoided it except where the oral evidence was crucial. Views were further acts by Hooke in public places. Even coffee houses were places where he worked: 'At Hublons View 1 G⊙. I paid J. Oliv. for yesterdays View 10. Viewed it again with J.O. I drew Report at Jonathans, we both signed it.'[167] Hooke seems to have had little trouble in persuading neighbours to agree on a settlement, or in making a citizen pay to have a defective wall taken down and rebuilt properly if he decided it was against the Acts or otherwise irregular, or of poor workmanship, or made from unsuitable materials. Although the City legally decided such things, they nearly always formally accepted the Surveyors' recommendations and ordered them to be carried out. The process worked well. Despite some obvious gaps in the archives, it has been estimated that Hooke made about 550 views of disputes between his first on 13 March 1668 and the end of 1674. Thereafter he made only a few views each year, but no report of a view by Hooke dated later than 1693 has been found.[168]

Negotiable fees were paid by citizens directly to the Surveyors for staking out and certifying foundations, measuring and certifying areas of lost ground, and viewing disputes. The main evidence for assessing how much Hooke received in fees for these services comes from his diaries, which on almost every page include notes of sums of money, usually associated with the name of a person or a place, but it is often difficult to decide if he is referring to a foundation certificate, an area certificate, a view, or some other service unconnected with his official City office.

Despite these difficulties, the loss of his survey books and the fact that he did not start to keep a diary until most of the rebuilding work was over, it has been possible to correlate enough evidence from his diaries and the City archives to make reasonable estimates of how much he was paid in fees for his duties as Surveyor up to the end of 1674, when most of the private rebuilding had been completed. He received about £1,000 for his foundation certificates, about £280 for his area certificates, and about £275 for his views.[169] From 1675 until the 1690s he continued to receive fees for his duties as Surveyor, but with greatly reduced frequency, amounting to at least £129.[170] In all he received at least an estimated £1,684 in fees from London's citizens for his work as City Surveyor, in addition to his salary, which is discussed below.

With most of the private rebuilding now completed, Hooke's surveying activities for the City in connection with his fourth task continued unabated: viewing and reporting to the City Lands Committee on contractors' estimates, bills of quantities, and the quality of workmanship and materials in the City's own secular rebuilding programme. He continued viewing with John Oliver, but was often called upon by the City Lands Committee as its technical consultant on major projects, many of which were specified in the Rebuilding Acts. The City relied on Hooke's authority in all technical matters, but the extent of his contribution has not been generally recognized. Reddaway pays even less attention to what Hooke did in rebuilding the City than to the work of Peter Mills and John Oliver, about whom he frankly admits that he knew 'so little that to set it down is to risk disparaging them'.[171] Nevertheless, Reddaway gives an unsurpassed account of the costs, difficulties, and ultimate failure of the two most ambitious projects in the rebuilding programme: the transformation of the squalid Fleet Ditch into a canalized tidal river, navigable as far as Holborn Bridge, flanked by wharves with storage cellars beneath and prestigious buildings behind; and the construction of a new broad quay with elegant buildings along the north bank of the Thames in place of the cramped alleyways, broken wharves, and foul detritus.[172]

Wren for the King and Hooke for the City worked together on the Fleet River. They were confronted by natural forces and pressures from the groundwater in the valley sides much greater than they had experienced. Several redesigns of the canal walls and increasingly expensive materials and methods were necessary to prevent collapse. From May 1672 to September 1674 Hooke and Wren held site meetings with contractors and the City, where they saw evidence of both progress and failure and debated what should be done. Wren left Hooke to deal with the contractors and write reports to the City, but Wren's endorsements of a succession of proposed changes in design and alarming increases in costs reassured the City that more money had to be spent if the project were to be completed. However, Reddaway's statement that 'the Fleet scheme is of unique interest as a part of the general secular rebuilding programme which can be indisputably attributed to Wren'[173] is true as far as it goes, but the scheme can also be indisputably attributed to Hooke. Although the City authorities were almost obsequious at times in their manner of addressing Wren and generous in the gratuities they awarded him for his advice, only Hooke from the City side had an intellectual ability and technical expertise that came close to matching Wren's. Moreover, Wren trusted Hooke to see through to completion what they decided together ought to be done and redone.

But a third figure was equally important in seeing that the Fleet River scheme was completed. In October 1674 Hooke referred to him as 'my good and sure freind'.[174] A leading City Whig, Sir John Lawrence was a generally respected City man who had courageously remained at his post as Lord Mayor during the Great Plague. He had supported Hooke in the past when he approved on behalf of the City Hooke's layout plan for rebuilding and when he chaired the City Side which rectified an earlier mistake and appointed Hooke Gresham Professor of Geometry. In 1673 he was elected to the Royal Society; later he was elected three times to its Council.[175] He and Hooke met socially, usually on a Saturday over dinner at Lawrence's house. As Hooke and Wren decided on what should be done at the Fleet River, so Hooke and Lawrence planned how to persuade the City Lands Committee to pay for it in the face of seemingly endless setbacks, revisions, and alarming increases in costs, which finally amounted to an astonishing £51,307 6s. 2d. (more than twice the money spent by the City from the Coal Tax for the whole of 1671),[176] but the scheme was completed successfully to the great benefit of people living nearby.[177] A detailed examination of evidence about Hooke's activities at a crucial stage in the construction in late 1672 reveals him rushing to and from the works at the Fleet, assessing the workmanship, checking the contractors' bills, and reporting formally in writing and in person to the City Lands Committee (sometimes as late as 8 p.m.) and informally at meetings with Wren and Lawrence separately and together at various places throughout the city.[178] The realization of the Fleet River scheme can be confidently attributed to Christopher Wren, Sir John Lawrence, and Hooke.[179]

The second great project on which Hooke worked on behalf of the City was the making of a new broad quay along the north bank of the Thames, stretching from the Tower upstream to the Temple, with rows of elegant south-facing houses to the rear overlooking the Thames.

Intended to replace decrepit wharves, decaying warehouses, narrow overcrowded alleyways, and filthy laystalls (where dung and refuse were stored for collection), the project ultimately failed for several reasons, all connected, as Reddaway has shown, with the rights of the tenants to continue trading and the lack of money to buy their ground and pay compensation for loss of trade.[180] Hooke was caught up in the disputes between the Lord Chancellor, the City, and the wharfingers, spending many hours in fruitless attempts to make progress. He supervised surveying for large-scale plans which showed existing and proposed shorelines. He visited wharfingers, warning them, as directed by the City and the Lord Chancellor, to take down their buildings. He discussed with Wren proposals for the work and reported them to the City Lands Committee, which sometimes needed the King's Commissioner to endorse what was being proposed, as it did with the Fleet River project. Again, Hooke's closeness to Wren enabled him in such cases to draft the report in such a way that Wren would agree to sign it.[181]

There were many other schemes that demanded Hooke's attention, less grandiose than the Fleet Channel and Thames Quay, but which brought much-needed improvements to the appearance of the city and increased the general well-being of its citizens. He was often taken along by the Alderman and Deputies of a Ward to view a technical problem with sewers, which was their responsibility to investigate and report on rather than Hooke's. Several examples can be found in the archives of reports written by Hooke but not signed by him because he was not officially ordered to undertake the work (Fig. 13). Hardly any new works in the city took place without Hooke's participation. He worked on gateways, particularly Newgate and Moorgate,[182] gave advice to the Commisioners for Sewers on gradients and routes for sewers and conduits and paving the streets,[183] decided on the locations for public latrines ('houses of common easement'), laystalls and slaughterhouses,[184] and had much to do with the preparation of ground for new markets,[185] including the supervision and approval of William Leybourne's surveys.[186] The Monument, long attributed to Wren, is now seen to have been designed in its final form by Hooke and built under his close supervision.[187]

Salaries for the City's Surveyors of New Buildings were paid from the Foundation Cash. Initially Peter Mills and Hooke were paid £100 per annum. In contrast to his salary payments by the Royal Society, the City paid Hooke from the time he began work for them (within a month of the outbreak of the Fire) even though he was not officially appointed Surveyor until six months later. He signed for his first payment of £25 (half the £50 due for work from Michaelmas 1666 to Lady Day 1667) on

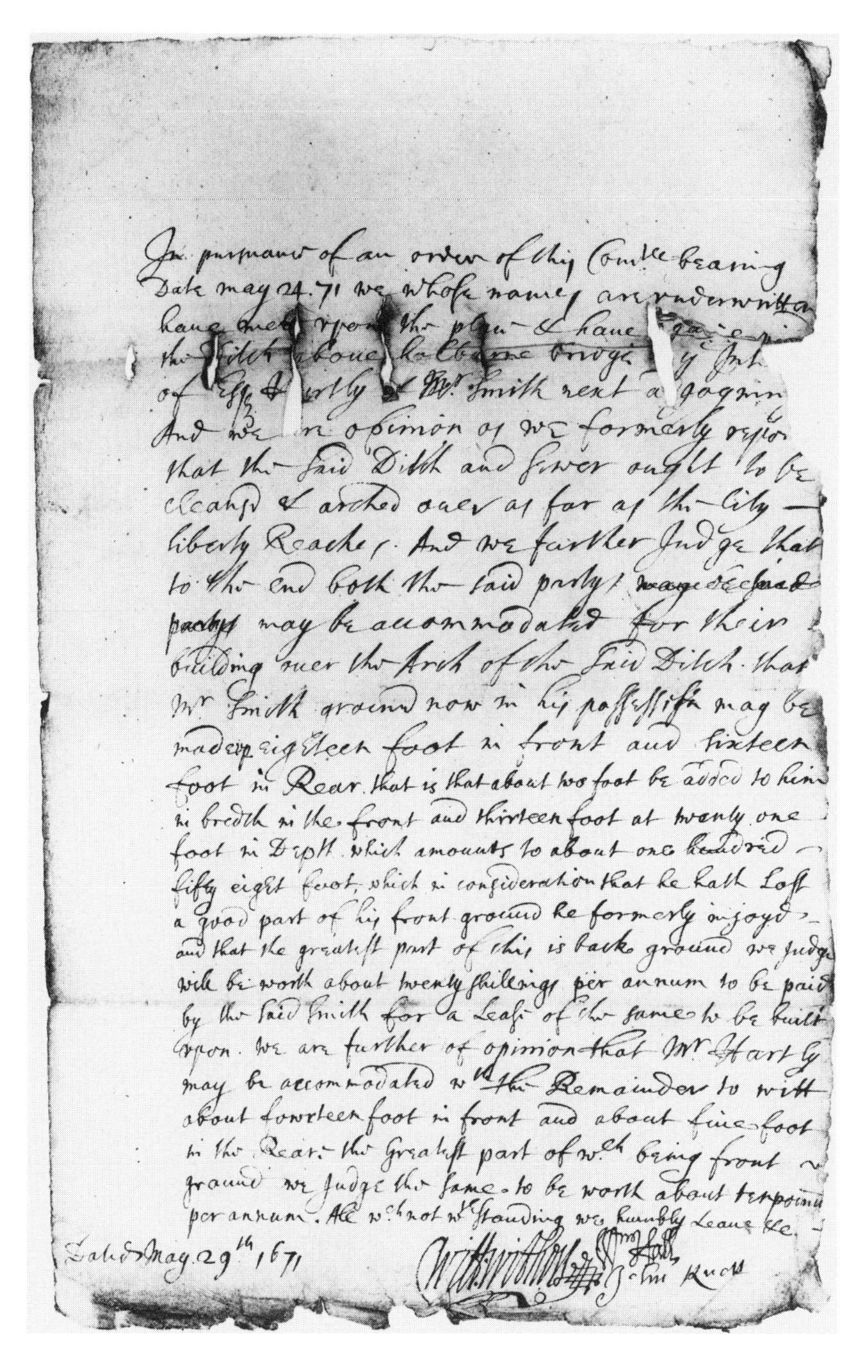

In pursuance of an order of this Comittee bearing
Date May 24. 71 we whose names are underwritten
have met upon the place & have [illegible]
the Ditch above holborne bridge [illegible]
of [illegible] Hartly & Mr Smith next adjoyning
And we [illegible] opinion as we formerly repor[t]
that the said Ditch and sewer ought to be
cleansd & arched over as far as the City
liberty Reaches. And we farther Judge that
to the end both the said partys ~~[illegible]~~
~~partys~~ may be accommodated for their
building over the Arch of the said Ditch that
Mr Smith ground now in his possession may be
made up eighteen foot in front and sixteen
foot in Rear. that is that about two foot be added to him
in bredth in the front and thirteen foot at twenty one
foot in Depth. which amounts to about one hundred
fifty eight foot, which in consideration that he hath Lost
a good part of his front ground he formerly injoyd
and that the greatest part of this is back ground we judge
will be worth about twenty shillings per annum to be paid
by the said Smith for a Lease of the same to be built
upon. We are further of opinion that Mr Hartly
may be accommodated wth the Remainder to witt
about fowrteen foot in front and about five foot
in the Rear. the Greatest part of wch being front
ground we judge the same to be worth about ten pounds
per annum. All wch not wthstanding we humbly Leave &c.

Dated May 29th 1671

[illegible] John Knott

Fig. 13 A report of a view of a dispute written by Hooke, but not signed by him. Matters in dispute which involved the City's own land or property, or which were not strictly to do with the new building, such as water courses and drainage, were settled by nominated Ward Aldermen and Deputies, but they often took one of the City Surveyors along for their technical advice. (City Lands Committee Papers MS 25, Corporation of London Records Office.)

9 July 1667. He signed for the balance of £25 three weeks later.[188] Surveyors' salaries were then increased to £150 (which might have persuaded Oliver finally to accept an appointment as Surveyor in January 1668), paid half-yearly until Lady Day 1668 and quarterly thereafter (Fig. 14) until Lady Day 1673, when Hooke and John Oliver (Peter Mills

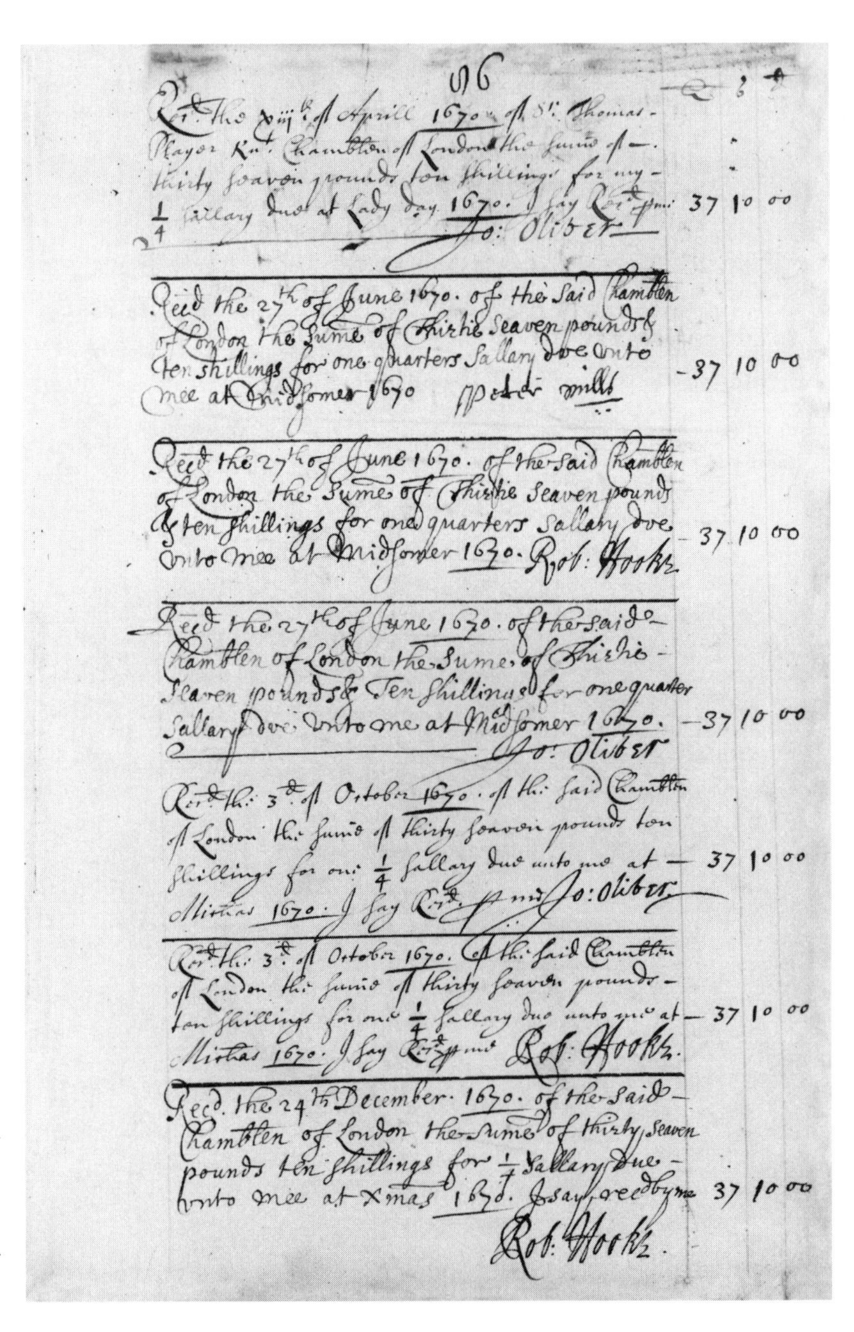
86

Recd. the xiijth of Aprill 1670 of Sr. Thomas Player Knt. Chamblen of London the summe of thirty seaven pounds ten shillings for my 1/4 sallary due at Lady day 1670. I say Recd p me — 37 10 00
Jo: Oliver

Recd the 27th of June 1670. of the Said Chamblen of London the Sume of Thirtie Seaven pounds & ten shillings for one quarters Sallary due unto mee at Midsomer 1670 Peter Mills — 37 10 00

Recd the 27th of June 1670. of the said Chamblen of London the Sume of Thirtie Seaven pounds & ten shillings for one quarters Sallary due unto mee at Midsomer 1670. Rob: Hooke. — 37 10 00

Recd the 27th of June 1670. of the said Chamblen of London the Sume of Thirtie Seaven pounds & Ten shillings for one quarter Sallary due unto me at Midsomer 1670. — 37 10 00
Jo: Oliver

Recd the 3d of October 1670. of the said Chamblen of London the summe of thirty seaven pounds ten shillings for one 1/4 sallary due unto me at Michas 1670. I say Recd p me Jo: Oliver — 37 10 00

Recd the 3d of October 1670. of the said Chamblen of London the summe of thirty seaven pounds ten shillings for one 1/4 sallary due unto me at Michas 1670. I say Recd p me Rob: Hooke. — 37 10 00

Recd. the 24th December 1670. of the said Chamblen of London the sume of thirty Seaven pounds ten shillings for 1/4 Sallary due unto mee at Xmas 1670. I say recd by me — 37 10 00
Rob: Hooke.

Fig. 14 The signatures of Mills, Hooke, and Oliver for receipt of their quarterly salaries as City Surveyors. (Ex-Guildhall MS 275 folio 86r, Corporation of London Records Office.)

had died in August or September 1670) received only £15 each. By that time the supply of Foundation Cash had almost dried up; surveys of only thirty foundations (equivalent to an income to the City of £10) had been requested in the first quarter of 1673.[189] No further payments of Surveyors' salaries were made from the Foundation Cash, but Hooke records 'Received of chamber £135 which with £15 formerly Receivd made £150 for last years Salary ending at Xtmas'.[190] From Michaelmas 1666 to Christmas 1673 Hooke had been paid £1,062 10s. in salary from the City. For the next two years Hooke and John Oliver received no regular salary payments. Instead they were paid different and irregular gratuities of varying amounts.[191] At its December meeting in each year, the City Lands Committee ordered salary payments to the Surveyors, so Hooke and Oliver waited on the evening of 20 December 1676, hoping to hear from the Committee about their arrears. They waited until 8 p.m. when the meeting finished, only to find that the Committee had forgotten them. With the meeting now over, Hooke and Oliver immediately raised the matter of their salaries with the City Chamberlain and some of the Committee, who agreed that payments of £50 should be ordered, but it is not clear whether the payments were made.[192] On the last day of 1676 Hooke made a list of his debtors and creditors. He was owed £1,300, which included his Surveyor's salary of £450 for the three years 1674–6. His seven creditors taken together were owed about £10, but in the middle of the list of creditors' names Hooke touchingly adds 'Much love to all my friends I owe.'[193] Another fleeting glimpse of his private thoughts comes in the same place where he writes, after working out the great difference between the money he owes to others and the money others owe to him, 'I owe none els a farthing to my knowledge,' which remark hints at a sense of unfairness, or even injustice, to himself.

Later that year Hooke and John Oliver met over threepenny chocolate at Jonathan's coffee house. Hooke spoke about their salary arrears and Oliver said he would join him in taking the matter to the City,[194] but the City would henceforth pay its Surveyors only gratuities. The need to allow private citizens to rebuild quickly had been met. There was no need now to pay salaries to the Surveyors. Hooke and Oliver continued to view buildings which were in dispute in the context of the Rebuilding Acts, but there were only a few of them each year, and the traditional craftsmen appointed as City Viewers were gradually taking up their former role. The City had always paid gratuities for particular services (Wren's were generous) and Hooke and Oliver were now treated in the same way, but much less generously. After payments of £40 for 1677, £15 for 1678, and £20 in 1679,[195] Hooke received irregular and infrequent payments from the City Lands Committee[196] which were almost always £5, enough at least to buy a new velvet coat once a year.[197] In some years he was paid nothing. Following his last salary payment to the end of 1673, he received no more than an estimated £350. His main work for the City was finished, but he continued to issue foundation and area certificates, take views, and give advice on building and construction whenever called upon by the City to do so.

He was also heavily involved with Wren in rebuilding London's parish churches, for which, by Wren's warrant, he was paid fees out of the money received by the City from the Coal Tax. Between 1671 and 1693 Hooke was paid a total of £2,820 for this service.[198] The annual payments, recorded in the account of salaries paid to officers and servants employed in building the parish churches, varied from £250 (in 1671, 1672, and 1683) to nothing (in 1689 and 1690). The first important account of Hooke's architectural work was published more than sixty years ago,[199] but in recent years scholars have added to our understanding of Hooke's architecture. Most of this new work is cited by Alison Stoesser-Johnston in her study of Dutch influence on Hooke's architecture[200] and by Paul Jeffery in his book on the City churches.[201] An explanation of why Wren authorized such a huge sum of money to be paid in fees to Hooke emerges from Paul Jeffery's account. The Commissioners for rebuilding the city parish churches (the Archbishop of Canterbury, the Bishop of London, and the Lord Mayor of the City) were appointed by Parliament in the 1670 Rebuilding Act. They required Wren, Hooke, and Edward Woodroofe[202] to undertake the rebuilding and repairs. First-hand evidence from Hooke's diary, parish records, and architectural drawings shows that he was often engaged in rebuilding or repairing many of the churches, but it has so far been difficult to say with confidence exactly what he did in connection with any particular church. Accumulating evidence, however, does point to Hooke having done more work at the building sites, supervising construction and certifying quantities and workmanship, than at the drawing board designing new churches.[203] He is now confidently said to have had an important role in the design of at least three of the existing city churches: St Anne and St Agnes (Gresham Street), St Benet Paul's Wharf (Queen Victoria Street), and St Edmund the King and Martyr (Lombard Street).[204]

Hooke also obtained several private commissions. Some of the more important of his buildings in London for individual or institutional clients, but which are now lost, were Aske's Hospital, Hoxton (1690s) rebuilt in 1826, Bedlam Hospital, Moorfields (1675–6) demolished in 1815–16, Bridewell Hospital (1671–8) demolished in 1862, Merchant Taylors' School, Suffolk Lane (1674–5) demolished in 1875, Montagu House, Bloomsbury (1675–9) gutted by fire in 1686, and the Royal College of Physicians, Warwick Lane (1672–8), where the theatre was demolished in 1866 and the rest destroyed by fire in 1879. A higher proportion of his buildings outside London remain, including Seth Ward's Almshouses at Buntingford, Hertfordshire (1689 or later), Ragley Hall, Warwickshire (1679–83), Ramsbury Manor, Wiltshire (1681–6), Shenfield Place, Essex

(1689), and, for Richard Busby, Willen Church, Buckinghamshire (1680; see Fig. 1).[205]

Evidence of his income from these architectural commissions is very sparse, compared with records of his salaries. Although he sometimes had to wait a few years before he received his fees in full for these tasks, they added significantly to his income. Even from the little evidence that has so far been found, it has been possible to show that he received over £250 from private commissions in the years 1670–4—a period when he was hardly engaged at all in private practice—and that thereafter he was paid a fee of at least £50 annually for each project he was engaged on.[206] He was also commissioned to advise on repairs and renovations at various places including Westminster Abbey, Lutton Church (now Sutton St Nicholas) in Lincolnshire, where Richard Busby was christened, and Sir Robert Southwell's Kings Weston estate near Bristol. Southwell, a statesman and diplomat elected President of the Royal Society in 1690, commissioned Hooke to undertake designs and tests of sluices and mills, for which he was paid at least twenty guineas.[207] Hooke's fees for his architectural and building commissions are difficult to assess from the little evidence available at present, but for the sake of the full picture they are estimated at £2,000.[208] Hooke's career earnings amounted to about £12,000.[209]

A Different Kind of Reckoning

Hooke's career was a virtuoso display of energy, genius, and probity across an astonishing variety of activities (Fig. 15). It is probable that his work for the Royal Society meant the most to him, yet it paid the least money. Only Hooke could have kept the Royal Society going during its difficult early days, at the same time dealing efficiently and promptly with the needs of thousands of London's citizens at a desperate time in their lives. In serving the City he showed great civic virtue. At a time when people were desperate to rebuild their homes and businesses, preferential treatment could readily have been bought and sold, but no evidence has been found that Hooke was anything other than scrupulously fair in all his dealings with the citizens of London and zealously methodical in keeping records of his multifarious acts. He performed his experiments in an optimistic pursuit of knowledge that he knew would be of great benefit to mankind, and in rebuilding London he worked with exceptional diligence and efficiency to improve the health and general well-being of its citizens. He was also one of the greatest experimental scientists of his time.

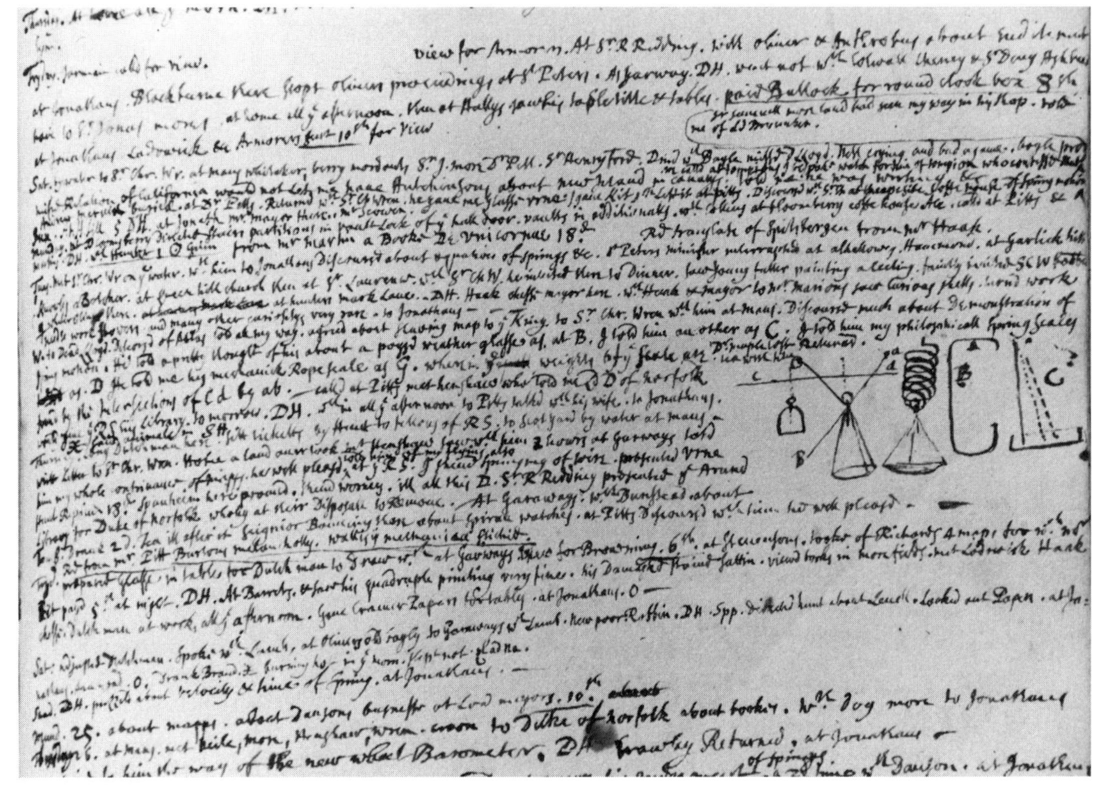

Fig. 15 Hooke's diary entry for 21 August 1678, showing some devices he discussed with Wren, including (left) a scheme by Wren for weighing by making use of the geometry of what we now call the triangle of forces and (second left) the spring Hooke was using to study velocities of oscillations. (MS 1758, Guildhall Library.)

Notes

1. Aubrey 1898, 409–16, and 2000, 393–7.
2. Hooke 1705, i–xxviii.
3. Hooke 1935, xiii–xxviii.
4. Nakajima 1994, 14–15 (before which it was generally believed that Hooke's family legacy was £100).
5. Aubrey 2000, 394.
6. Hooke 1705, xxvi.
7. Hooke 1705, iii.
8. Aubrey 2000, 394.
9. Nakajima 1994, 15.
10. Jardine, 2002, 14–23 (which acknowledges the archival work of Rachel Jardine).
11. Barker 1895, 95–9.
12. Barker 1895, 104–27.
13. Carleton 1965, 16.
14. Aubrey 2000, 395.
15. Hooke 1665, preface, sig. g1v.
16. Aubrey 2000, 395.
17. Aarsleff 1970–80, 361–81.
18. Hooke 1665, preface, sig. g2v.
19. Jim Bennett discusses below Hooke's optimism that the devices he conceived, but could not make because of the limitations of contemporary materials and methods, would one day be made.

20. In the 1680s, when Ward was Bishop of Salisbury, he commissioned Hooke to design some almshouses at Buntingford in Hertfordshire, Ward's birthplace.
21. Hooke 1705, iv.
22. Willis's father-in-law, Dr Thomas Fell, was rector at Freshwater when Hooke's father was appointed curate there in the 1620s (Jardine 2002, 116).
23. Hooke 1705, iii.
24. Shapin 1994, 355–407.
25. Davis 1994, 157–75.
26. On 14 February 1667, shortly after moving from Gresham College to Arundel House following the Great Fire, the Society changed its meeting day to Thursday (Birch 1756–7, ii, 149).
27. Birch 1756–7, i, 7 has 'forty shillings' but this was increased to £4.
28. Hunter 1989.
29. Michael Hunter shows clearly and in detail the severe financial constraints on the Society brought about by the failure of members to pay their dues (Hunter 1994c).
30. Hunter 1994c, 36.
31. Birch 1756–7, i, 21.
32. Birch 1756–7, i, 123.
33. Birch 1756–7, i, 124.
34. Birch 1756–7, i, 453.
35. Sir Thomas Gresham's original foundation made provision for seven Professors: Divinity, Astronomy, Geometry and Music (appointments made by the City of London); and Law, Physic and Rhetoric (appointments by the Mercers' Company). It was not until 1984 that an eighth Chair (Commerce) was established, sponsored by the Mercers' School Memorial Trust, appointments by the Mercers' Company in consultation with the trustees and the College Council (Chartres and Vermont 1998, 67).
36. Birch 1756–7, i, 503. But later, Sir John Cutler, in the 1680s, would claim that he also had intended to benefit the Royal Society.
37. Hunter 1989, 279–338.
38. The Royal Society Account Books give no indication that Hooke was paid any salary at all until the accounting period ending 10 April 1666 during which he was paid £20 (Royal Society Account Books I). This sum probably represents members' voluntary contributions to Hooke's salary referred to in Council's order of October 1664 to the Treasurer and which the President on 28 June 1665 was authorized to sign for (Royal Society Council Minutes (Copy) I, 105). In addition to the payment of £20, Hooke received in the same period two payments (of £3 and £3 6s.) but they were refunds of sums paid out by him on behalf of the Society. The Treasurer was authorized by Council to make the payment of £20 (Royal Society Council Minutes (Copy) I, 119, dated 4 July 1666). In the next accounting period, which ended on 5 November 1666, Hooke was paid £30 (Royal Society Account Books I). At this date the Society's accounts (of which Hooke was an auditor) show a deficit of £678 5s. 0d.
39. For details of how the Society arrived at this figure for the arrears of Hooke's salary, see Hunter 1989 and Cooper 1999, 233.
40. Royal Society Account Books I.
41. Royal Society Account Books I, II, and III, passim.
42. Shapin 1988a.
43. Iliffe 1995.
44. Birch 1756–7, i and ii.
45. Birch 1756–7, i, 311–21. For a recent meticulous account of the Royal Society's early problems see Hunter 1989 passim; Lyons 1944, 72–117 (an account of some of the Royal Society's more difficult years); Sprat 1667 (the first published history of the Royal Society); Gunther 1930a and b republishes much on Hooke extracted from Birch 1756–7 (but without specific references) together with a variety of transcripts, etc. from other sources.

46. Birch 1756–7, ii, 182–5.
47. Boyle 2001, iii, 332.
48. Hunter 1994c, 110–11.
49. Birch 1756–7, ii, 289.
50. Birch 1756–7, ii, 346 dated 11 February 1669, 354 dated 11 March 1669 (when Hooke attended the meeting, but instead of carrying out the experiments ordered he reported on microscopical observations of frogs he had taken upon himself), 383 dated 17 June 1669 (when Hooke was again present, but excused himself from performing the experiments ordered), 395 dated 22 July 1669 (when at the last meeting before the recess Hooke was ordered to perform in private all the experiments already committed to his care, but not yet performed, and report on them to the Society when it reconvened), 411 dated 9 December 1669 (when Hooke excused himself, saying that avocations of a public nature had prevented him from performing the experiments ordered), and 418 dated 10 February 1670 (when Hooke was absent).
51. Birch 1756–7, ii, 383.
52. An investment of £2860 in houses, or of £3740 in land, would be needed to meet the estimated annual costs of £500 (Hunter 1989, 194–5). Hooke's hitherto rather shadowy presence in the Society's attempts at financial and procedural reform at this time and in the 1680s has been revealed and discussed by Michael Hunter (ibid. 193–202, 232–9, and 242–4).
53. Birch 1756–7, ii, 469.
54. Hunter 1989, 190. Seth Ward, Bishop of Salisbury, taught Hooke astronomy at Christ Church, Oxford, and in the 1680s commissioned Hooke to build almshouses in his (Ward's) name at Buntingford in Hertfordshire.
55. Hooke had distrusted Oldenburg for a long time, thinking he traded in information and broke confidences. Such thoughts were behind the formation of the 'Philosophical Club' comprising a few active Royal Society members who met privately for open philosophical debate without fear that their ideas or theories would be passed on clandestinely. Hooke's elections to Council and to the office of Secretary took place at the Society's annual meeting held on St Andrew's Day, 30 November, the anniversary of the Society's foundation.
56. Abraham Hill, the Society's Treasurer, was one of the grandees in the Royal Society whom Hooke came to resent, as revealed in his diary entries relating to Hill's support of Nehemiah Grew rather than Hooke as the Secretary to sit at the officers' table and take notes at meetings:

> Hill slandered me to Sir J. Williamson [President] . . .
>
> Met Grew at Martins, and discoursd at Jonathan's. He would have been to sit at the table at the meeting. He was going to Hill slyly bragged of Brother. I told him of my former abuses etc. . . .
>
> This evening Hill principally and Louther plotted for Grew againste me at the Secretarys. *Vindica me Deus* . . .
>
> Grew placed at table to take notes. It seemed as if they would have me still curator. I stayed not at the Crown. I huffed at Hill at Jonathans . . .
>
> At the Councell . . . Grew to take notes also but I to draw them up. Sir J. Williamson wrote the Determinations of this Councell himself, was pleased with what I proposed about Experiments. Hill a dog for Grew. Hill and Louther slyly. Sir J. Louther wanted a Rat. The President kinder to me . . .
>
> A Hill affronted me at Crown. *Cave canem*

(Hooke 1935, 332 (4 December 1677), 333 (3, 12, and 13 December 1677), 335 (19 December 1677), and 338 (3 January 1678). The brother referred to was probably Grew's half-brother Henry Sampson, a successful London physician (LeFanu 1990, 1).

57. Hooke 1665, Preface, sig. a1r.

58. Birch 1756–7, iii, 501.
59. Birch 1756–7, iii, 514.
60. Birch 1756–7, iv, 91. Hooke received the payment of £40 on 20 January 1682 (Royal Society Account Books II).
61. Birch 1756–7, iv, 168.
62. Birch 1756–7, iv, 174, 196, and 260.
63. Birch 1756–7, iv, 188. They were, however, paid £20 each in cash (ibid., 245).
64. Hunter 1989, 331.
65. Royal Society Account Books III.
66. Birch 1756–7, ii, 489.
67. Royal Society Account Books III. In June 1687 Council reaffirmed its original order, and decided that his payment should receive precedence over others (Birch 1756–7, iv, 542) but the following month its members had the idea of paying Hooke (and Edmond Halley as Clerk) in copies of a Royal Society publication, *History of Fishes* (which was not selling well) in lieu of cash, at a rate of one book for £1. Hooke said he would think about the proposal for six months before letting Council know whether or not he was willing to accept the books in lieu (Birch 1756–7, iv, 545). He was paid in cash.
68. On 29 November 1687 Council ordered a payment of £22 10s. 0d. as Hooke's salary from Lady Day to Christmas 1686 (Birch 1756–7, iv, 554). The payment was made on 17 May 1688; another payment of £37 10s. 0d. for salary from Christmas 1686 to Lady Day 1687 was made on 5 September 1688 (Royal Society Account Books III). No further payments of salary for his work as Curator of Experiments after Lady Day 1687 have been found in the Treasurer's account books.
69. This sum comprises salary payments of £80 annually from Lady Day 1664 to 23 November 1664, and £30 annually thereafter until Lady Day 1688.
70. On 13 March 1691 Hooke intended to state his case for resuming his salaried curatorship to Council at its next meeting (Royal Society Council Minutes (Copy) II, 111) but does not seem to have done so.
71. On 12 April 1702 the Society agreed to pay him 'not more than £15' presumably for his part in support of the action taken by the Society to protect its interests in the Bill before Parliament which would allow the Gresham Committee to demolish its College. Hooke had presented to the House of Lords on 24 March 1702 objections to the Bill, which are described by Ian Adamson as 'hastily constructed, exaggerated, inaccurate and scrupulously avoid mentioning the Royal Society'. Hooke's ostensibly personal intervention delayed the passage of the first Bill. The second Bill failed. The Royal Society of its own accord in 1711 moved out from a dilapidated Gresham College to a house in Crane Court (Adamson 1978, 10–15).
72. Royal Society Account Books III.
73. Hunter 1989, 329.
74. Hunter 1989, 322 and 324.
75. Hooke 1705, xxv. According to Waller, Hooke recorded these words in a diary, which is now lost.
76. A few of Hooke's Cutlerian Lectures were published individually in the 1670s and republished as *Lectiones Cutlerianae* in 1679. As Michael Hunter shows below, they are important for an understanding of Hooke's natural philosophy, but some are lost and others survive in manuscript form in various places and remain unpublished (Hunter 1989, 298–305).
77. Referred to hereafter as 'the Gresham Trustees'.
78. Referred to hereafter as 'the Gresham sub-committee'.
79. Mercers' Company Gresham Repertory II, 208.
80. Sir Richard Browne was a coal merchant and had been Lord Mayor 1660–1 and MP for various constituencies (Woodhead 1965, 39–40). No clear reason why he persuaded Arthur Dacres to withdraw his application has been found, but it seems as if the

certificates produced at the first meeting by Dacres were inferior to those of Isaac Barrow, although certificates produced at the second meeting by Dacres (when Hooke was a candidate) were more favourable to him than his earlier ones. Barrow was chosen by unanimous consent (Mercers' Company Gresham Repertory II, 209).

81. John Wilkins had much to do with Isaac Barrow's appointment at Cambridge, from which he was to resign in 1669 in favour of his brilliant pupil, Isaac Newton (Ward 1740, 160–1).
82. A footnote by Thomas Birch is explicit in stating that the Society was eager for Hooke to be appointed Gresham Professor of Geometry (Birch 1756–7, i, 435). It is probable that John Wilkins and other members of the Society spoke in Hooke's favour.
83. Mercers' Company Gresham Repertory II, 215.
84. Birch 1756–7, i, 435.
85. Mercers' Company Gresham Repertory II, 217.
86. Mercers' Company Gresham Repertory II, 218. Ian Adamson (1978, 3–4) has written that the votes were equal at five each and the Lord Mayor gave his casting vote, which is incorrect, and Patri Pugliese (1982, 3–4) is also incorrect in writing that the votes were five for Hooke and five for Dacres, including the Lord Mayor who then declared Dacres elected. The misdemeanour was greater than they have described. The Lord Mayor was not legally entitled to be present, but he took control and in giving (i.e. announcing) the result of the vote (which was five to four in favour of Hooke) he declared Dacres elected.
87. PRO C7/564/29, item 2, discussed by Hunter (1989, 287) who earlier (ibid., 284) points out that Pepys in his diary noted that Sir John Cutler and John Graunt (a member of the Royal Society and friend and colleague of Sir William Petty) were often seen together in coffee-houses. Hunter goes on to say that Graunt and Petty had a leading role in the negotiations between Cutler and the Royal Society.
88. Woodhead 1965, 71.
89. Ward 1740, 19–25.
90. Chartres and Vermont 1998, and Featherstone 1952, give details of the complex history of the implementation of Sir Thomas Gresham's will.
91. Ordinances and Agreements dated 16 January 1597 between the Joint Grand Gresham Committee and the first Professors can be found in Ward 1740, iii–viii.
92. Mercers' Company Gresham Repertory III, 133.
93. Adamson 1978, 5.
94. Sir Andrew King was a London merchant, a member of the Royal Society (Hunter 1994c, 158–9), and a relative of a Gresham Professor of Rhetoric, John King. He also rented the lodgings of Thomas Baines, Professor of Music, when Baines was absent overseas. He died in the College in 1679 (Ward 1740, 329). From time to time Hooke dined with King in the College, once on beef and goose. Hooke noted in January 1673 that King's chimney 'in the common kitchen' was brought down in a storm. The following April Hooke set up some cupboards for King (Hooke 1935, 8, 128, 20, 41).
95. Adamson 1978, 5.
96. Adamson 1978, 6.
97. Ward 1740, 65–70.
98. Boyle 2001, ii, 342–4. Walter Pope, FRS, John Wilkins's half-brother and formerly a scholar at Wadham, who succeeded Christopher Wren as Gresham Professor of Astronomy, was on tour in Italy at the time (Ward 1740, 112).
99. Abraham Hill signed for Pope's £25 salary on three occasions: 26 November 1663 for Lady Day to Michaelmas 1663; 12 May 1664 for Michaelmas 1663 to Lady Day 1664; and 15 November 1664 for Lady Day to Michaelmas 1664 (Corporation of London Records Office, Gresham College and Royal Exchange Acquittance Books I, folios 51r and v and 65r). In signing for the first payment Hill anticipated his appointment as Treasurer by four days (Birch 1756–7, i, 337).

100. Ward 1740, 112.
101. Corporation of London Records Office, Gresham College and Royal Exchange Acquittance Books I, folio 66r.
102. Ward 1740, xv–xvi.
103. The day and time of the Royal Society meetings were changed to Wednesday (the day of their earliest meetings) at 4 p.m., beginning on 12 January 1681 (Birch 1756–7, iv, 60, dated 8 December 1680).
104. Hooke 1705, 65–202. Most of these lectures or discourses are identifiable as Gresham Geometry Lectures, but some might be transcripts of Hooke's writings on related topics which he did not read as Gresham Lectures.
105. Hunter 1989, 279–338.
106. Adamson 1978, 6. Some further comments by Hooke on the absence of an audience for his Gresham Lectures are:

> Noe auditory came either afternoon or morning soe I read not.
> Noe Lecture. I having stayd till past 3. Harry spoke to one to come earlier as I also did.
> stayed till 2¾ noe company for Lecture.
> None came till past 3 for Lecture then two which grumbled.
> 2 young fellows inquired for Lecture at 3.
> Baker and another to hear Lecture after 3, Grace spoke to them.
> Noe auditory, m[orning] foggy . . . noe auditory from 2 to 3.
> Noe auditors [morning] . . . HH [Harry Hunt, Operator] pp. tea. He attended hall till 3: noe auditor came.

(Hooke 1935, 47 (12 June 1673), 226 (13 April 1676), 257 (16 November 1676), 258 (23 November 1676), 354 (18 April 1678), and 413 (22 May 1679). Gunther (ed.) 1935, 72 (15 November 1688) and 104 (18 April 1689)).

107. Gunther (ed.) 1935, 118 (2 May 1689). Edward Paget was a member of the Royal Society and mathematics master at Christ's Hospital (Hunter 1994c, 208–9).
108. Hooke 1935, 236 (8 June 1676).
109. Hooke 1935, 446 (17 June 1680).
110. Hooke 1935, 382 (31 October 1678).
111. 'Espinasse 1956, ch. 7.
112. Gresham College and Royal Exchange Acquittance Books III, folio 8v.
113. For accounts of the Fire and its aftermath see, for example, Bell 1923, Reddaway 1951, Milne 1986, and Porter 1996.
114. Porter 1996, 60.
115. Cooper 1997; 1998a, b.
116. Perks 1922, 143–52.
117. 'The City' is used henceforth to mean the men, committees, courts, and officers of the City of London and 'the city' is used to mean the geographical City and Liberties of London.
118. Corporation of London Records Office Repertory 71, folios 168v–9v.
119. On 4 July 1667 the Society decided to move its meetings to Arundel House, returning to Gresham College in 1673 (Adamson 1978, 2).
120. Hooke was probably allowed to remain because the City had already decided to use him in the task of rebuilding, having approved his layout plan only a few days earlier (Birch 1756–7, ii, 115).
121. Rasmussen 1937, 93–114; Porter 1996, 92–116.
122. Birch 1756–7, ii, 115.
123. Hooke 1705, xii–xiii.
124. Reddaway 1951, 53.

125. The hugely over-ambitious survey was the king's idea. The intention was to make a new and accurate large-scale plot showing all the property boundaries and use it as a basis for planning and administering the new city. Information about each parcel, or toft, such as ownership, rents, occupier, and to whom the inheritance or reversion applied was to be collected and entered into registers. If there had been enough time to complete the survey, London would have had a parcel-based land information system, but there was not enough time even to clear away the rubble so that the streets and old foundations could be seen and survey measurements taken.
126. Corporation of London Records Office Journal 46, folio 123r dated 4 October 1666. Christopher Wren at the time was Savilian Professor of Astronomy at Oxford, beginning to gain a reputation as an architect. Hugh May was the King's Paymaster (later Controller) of Works and had dealt with the repair of the royal palaces after the Restoration. Roger Pratt was a gentleman architect, having built Lord Clarendon's residence in Piccadilly and other grand mansions. Peter Mills had been City Bricklayer, City Viewer, and was the present City Surveyor, whose layout plan had been rejected by the City in favour of Hooke's. Edward Jerman (sometimes 'Jarman' or 'Jarmyn') had been City Carpenter, City Viewer, and had resigned as City Surveyor in 1657 to return to lucrative private practice.
127. The first of Hooke's experiments on breaking wood (an early example of the engineering discipline 'strength of materials') took place at the Royal Society on 8 June 1664 (Birch 1756–7, i, 435–6).
128. Hull 1997. Hooke's illustration of Kettering-stone for *Micrographia*, published in early 1665, and Wren's use of the material for Pembroke College Chapel (1663–5) are evidence of the close relationship between the two men before the Fire.
129. Royal Society Classified Papers XX, 29 dated 8 June 1664, the day the Royal Society Council (at a meeting immediately preceding the meeting of the Society when Hooke presented his report on the thunderstorm) heard a report that Dacres's earlier election as Gresham Professor of Geometry in preference to Hooke was irregular.
130. Reddaway 1951, 58.
131. Jardine 2002, 259.
132. See Bennett 1982 for Wren and Hooke in science and Jardine 2002 for Wren and Hooke in architecture and building.
133. Corporation of London Records Office Journal 46, folio 129r (31 October 1666).
134. At the same meeting, the Court of Common Council considered a Parliamentary Bill on the making of bricks and lime. No evidence has been found to link Hooke or anyone else in the Royal Society with this Bill, but it is worth noting that on the afternoon of the same day, at a meeting of the Royal Society when they were discussing bricks,

    > Mr Hooke took notice, that those earths, which will vitrify, make the more lasting bricks.
    >
    > It was ordered, that Mr Hooke should make trials of several earths by burning them in a wind furnace, to see, which kind would yield the best brick.
    >
    > (Birch 1756–7, i, 118–19)

135. Saunders 1997, 128–34.
136. See n.71.
137. *An Act for the Rebuilding of the City of London*. 18 and 19 Charles II, c.8. Sections IV and VIII of the Act govern the Surveyors' appointment and duties (Jones and Reddaway 1967, xiv–xv).
138. Corporation of London Records Office Journal 46, folio 146v (25 February 1667).
139. Corporation of London Records Office Journal 46, folio 147v (13 March 1667). The map was probably a version of John Leake's manuscript compilation (British Library, Add.

MS 5415 E.1) from the six surveyors' original sheets. It shows the proposed widths of streets and intended new works, such as King Street and the Fleet Channel. The original six manuscripts have not survived.

140. Corporation of London Records Office Journal 46, folios 151r–2r. Two hundred copies were printed by James Flesher, the City's Printer (Corporation of London Records Office Printed Document 10.54(L)).
141. John Oliver had a varied and interesting career, but it has not yet been examined in detail. Master of the Glaziers' Company, surveyor (he became assistant surveyor at St Paul's following the death of Edward Woodroofe), and architect, he also dealt in real estate and was a conscientious municipal councillor and parish citizen (Jones and Reddaway 1967, xxx–xxxi, and Lang 1956, 98).
142. Corporation of London Records Office Repertory 72, folio 81v dated 20 March 1667.
143. Birch 1756–7, ii, 160.
144. Peter Mills wrote in his survey book 'Wee began to stake out the streets in ffleet street the the 27th. of March 1667' (Guildhall Library, MS 84.1, folio 22r, reproduced in London Topographical Society 1964).
145. Corporation of London Records Office ex-Guildhall Library MS 322/3.
146. Corporation of London Records Office ex-Guildhall Library MS 322/9 is the latest account of the Clerk of Works for expenses incurred in staking out the streets that has been found. It is for the seven days ending Monday, 26 May 1667.
147. Corporation of London Records Office Journal 46, folios 151r–2r dated 29 April 1667. Two hundred copies were printed by James Flesher, the City's Printer (Corporation of London Records Office Printed Document 10.54(L)).
148. This money, known as the 'Foundation Cash', was used to pay Surveyors' salaries and some other costs of rebuilding.
149. Guildhall Library MS 84.1, folios 16–18, reproduced in London Topographical Society 1964, are entries in Mills's survey books of surveys by Hooke and Oliver.
150. Corporation of London Records Office Repertory 73, folio 62r, dated 28 January 1668.
151. Guildhall Library MS 84.2, folio 86v, reproduced in London Topographical Society 1962a.
152. Corporation of London Records Office ex-Guildhall Library MS 276, folio 86v bears the signature of one of Mills's executors for receiving Mills's quarter salary due at Michaelmas 1670. The date of the signature is 7 October 1670.
153. Unlike area certificates and reports of views which had to be submitted to the City, foundation certificates were retained by individual citizens as evidence that their property had been staked out by a City Surveyor.
154. Guildhall Library MSS 84.1 and 84.2 are two volumes of transcripts of Mills's ten survey books and Guildhall Library MSS 84.3 and 84.4 are transcripts of Oliver's. Guildhall Library MS 84.5 is an index to the other four volumes.
155. London Topographical Society 1962a, b, c, 1964 and 1967.
156. Hooke's diary entry for 7 November 1677 includes 'with committee of City Lands all the afternoon till almost 7 at night, they enquired concerning my books and concerning Certificatts, would have my books Deliverd' (Hooke 1935, 326).
157. The first note in his manuscript diary (not included in Hooke 1935) is 'Memoranda begun: March 10 167$^1/_2$' (Guildhall Library MS 1758). Hooke might have decided to begin his diary when his work as City Surveyor diminished in 1672 and he no longer needed his survey books with him as regularly as in the past, when he nearly always had them at hand for jotting down memoranda.
158. Cooper 1997. See Birch 1756–7, ii, 383 for Hooke's failure to perform experiments for the Royal Society. Hooke nearly always used the mornings for his surveying until 1672, by which time the great urgency to rebuild had diminished.
159. The left-hand margin of this area certificate is evidence of Hooke's usage of mixed (decimal and duodecimal fraction—abbreviated to df here) notation and mental arithmetic in the calculation of area. First he writes down his measurement of the width

(20ft) of the first area of ground. Then he writes down 8.4 (df) which is the average of his two measurements of the east (8ft) and west (8ft 8in) depths of the first area. The area is found by multiplying 20 by 8.4 (df) which he does by first finding the product of 20 and .4 (df) which is 6.8 (df) so he writes down the .8 (df) and carries 6 in his mind. He then multiplies 20 by 8, adds the 6 he has carried, and writes 166. The first area is shown to be 166.8ft^2 (df). He proceeds similarly for the second area. The width of the ground is 41ft 10in which he writes as 40.10 (df). The two measured depths are 2ft 8in and 1ft 4in, so he writes the average as 2. To find the area he multiplies the 0.10 (df) by 2 and writes the result as 1.8 (df) then multiplies 41 by two and writes the result as 82. By addition the second area is found to be 83.8ft^2 (df). He then adds the two areas, first adding .8 (df) and .8 (df) to give 1.4 (df). He writes down .4 (df) and carries the 1. Then he adds 83 to 166, adds the 1 he has carried, and arrives at 250. The total area of ground lost is therefore 250.4ft^2 (df). Also shown on the manuscript is a clerk's note of £50 10s. 0d. which probably represents the compensation to paid by the City to Mr King at a rate of 4s. 0d. per square foot, allowing an exceptionally generous 10s. 0d. for the fractional part of a foot instead of the 1s. 8d. it warrants; generally the compensation was more accurately calculated than in this example.

160. 18 and 19 Charles II, c.8 of 8 February 1667, and 22 Charles II, c.11 of 11 April 1670.
161. Corporation of London Records Office Journal 46, folio 209r, dated 22 January 1668.
162. Corporation of London Records Office Comptroller's Deeds Box K.
163. Cooper 1998a; 1999, 51–63.
164. Disputes between landlords and tenants were settled in the Fire Courts (Jones (ed.) 1966).
165. The word 'view' has been used carelessly by writers on Hooke, even by those who are generally more scrupulous. Margaret 'Espinasse (1956, 86), the editors of Hooke's first diary (Hooke 1935, xxiv), and Steven Shapin (1989, 255) all use it indiscriminately in connection with Hooke's surveying, placing it in quotation marks without further explanation. A view was a specific activity (to visit a site, take evidence, and report on what was found) in connection either with the City's building works, where costs, materials and workmanship might be the subject of the report, or in connection with allegations of irregular private rebuilding or disputes between neighbours where infringement of the Rebuilding Acts was the subject. Hooke's foundation surveys and area certificates were neither views nor 'views'.
166. Generally two Surveyors were nominated by the Court of Aldermen to undertake each view. In the many manuscript reports to the Court of Aldermen that have been examined, not one that bears Hooke's signature has been found to be in the handwriting of either Mills or Oliver. By contrast, either Mills or Oliver (if nominated to view with Hooke) have signed all reports written by Hooke. This implies that Hooke took the lead in viewing disputes.
167. Gunther (ed.) 1935, 237 (6 May 1693). Hooke received a fee of one golden guinea for Hublon's view. At Jonathan's coffee-house in Exchange Alley Hooke wrote a report of a view they had begun the previous day. John Oliver was present and signed the report. Hooke then gave him ten shillings, Oliver's share of the fee for the earlier view. The diary entry reveals evidence that Hooke took the leading role when viewing with John Oliver.
168. Cooper 1998b; 1999, 68.
169. Cooper 1997; 1998a, b; and 1999, 145–6.
170. This estimate comprises £52 for foundation certificates, £22 for area certificates, and £55 for views (Cooper 1999, 192, 194, and 198). According to the City archives, Hooke's last foundation certificate is dated 29 June 1694 (Corporation of London Records Office ex-Guildhall Library MS 277 fol 14^r); his last area certificate is dated 11 March 1687 (Corporation of London Records Office Comptroller's Deeds Box K/H/72) and his

last view is dated 19 December 1693 (Corporation of London Records Office Misc. MS 93.79).

171. Reddaway 1940, 59.
172. Reddaway 1940, 200–43.
173. Reddaway 1940, 216.
174. Hooke 1935, 125.
175. Hunter 1994c, 190–1.
176. Reddaway 1940, 215 and 313.
177. The new channel was not cleansed regularly. Silt slowly accumulated until the upper reach above Fleet Bridge could no longer be used by boats. It was arched over in 1733 and the adjacent wharves were used as roads. Thirty years later the lower reaches were covered over. By the time the first Blackfriars Bridge was built in 1769, the navigable canal had reverted to its former stinking state and its replacement by a road was inevitable.
178. Cooper 1999, 78–86; Hooke 1935, 9 (2, 3, 7, 8, and 9 October 1672), 10 (12 and 16 October 1672), 11 (21, 22, 23, 25 October 1672), 12 (2, 5, and 9 November 1672), and 13 (12 and 14 November 1672); Corporation of London Records Office City Lands Committee Orders Books I, folios 47v (25 November 1670), 50r (8 February 1671), 50v–51r (1 March 1671), and 54r and v (22 March 1671); Corporation of London Records Office, City Lands Committee Papers, MSS 57, 1 (20 November 1672) and 65 (3 January 1673); and Corporation of London Records Office, City Lands Committee Minutes (Rough) for 16 and 23 October; and 13 and 20 November 1672.
179. This is not generally recognized. No mention of Sir John Lawrence has been found in Margaret 'Espinasse's biography of Hooke (1956). Reddaway's frank admission of uncertainty about Hooke's role in canalizing the Fleet River is based on his acceptance of only two of Hooke's diary entries as being relevant (Reddaway 1940, 216 n.6) whereas many others, when examined with contemporary City records (a sample of which are listed above), leave little doubt about what Hooke did and the importance of his role.
180. Reddaway 1940, 221–43.
181. For example: Cooper 1999, 86–8; Corporation of London Records Office, City Lands Committee Orders Books I, folios 55r, 56v, 59v, and 60r; Hooke 1935, 12 (9 November 1672), 44 (21 May 1673), 45 (26, 31 May, and 1 June 1673), 58 (29 August and 3 September 1673), 59 (4 and 5 September 1673), and 64 (8 October 1673).
182. For example: Cooper 1999, 92–4; Hooke 1940, 5 (14 August 1672), 47 (13 June 1673); Corporation of London Records Office, City Lands Committee Minutes (Rough) for 17 July 1672, 14 and 27 August 1672, 5 and 11 September 1672, 9 and 16 October 1672, and 15 and 29 January 1673; Corporation of London Records Office, City Lands Committee Orders Books II, folios 27v (17 July 1672) and 30r (14 August 1672).
183. For example: Cooper 1999, 92–3; Hooke 1935, 6 (11 September 1672); Corporation of London Records Office Repertory 72, folio 108r (16 May 1667); Corporation of London Records Office, Printed Document 10.117(L) dated 8 July 1667; Corporation of London Records Office, City Lands Committee Orders Books I, folios 54v (22 March 1671), 55r (3 May 1671), 56r (10 May 1671) II, folios 15r (30 April 1672), 29r (24 July 1672); Corporation of London Records Office, City Lands Committee Papers, MS 25 (29 May 1671) (Fig: manuscript written by Hooke, not signed by him); Corporation of London Records Office, City Lands Committee Minutes (Rough) dated 11 September 1672.
184. For example: Cooper 1999, 93–4; Hooke 1935, 11 (23 October 1672), 12 (6 and 8 November 1672); Corporation of London Records Office, City Lands Committee Orders Books I folios 43v (10 August 1670) and 66r and v (4 October 1671); Corporation of London Records Office, City Lands Committee Minutes (Rough) dated 23 October 1672 and 6 November 1672; Corporation of London Records Office Repertory 76 folio 15v (22 November 1670).
185. For example: Cooper 1999, 94–7; Hooke 1935, 14 (20 November 1672), 27 (7 February 1673), 45 (29 May 1673), and 61 (23 September 1673); Corporation of London Records

Office, City Lands Committee Minutes (Rough) dated 28 April 1669, 11 May 1669, 2 June 1669, 1 September 1669, 6 October 1669, and 10 February 1670; Corporation of London Records Office, City Lands Committee Papers, MS 7 (26 May 1669).

186. Masters 1974 gives descriptions of the provenance of these plans and reproduces them at reduced scales.
187. For example: Cooper 1999, 88–90; Hooke 1935, 11 (22 October 1672), 54 (8 August 1673), 59 (11 September 1673), 66 (19 October 1673), 93 (28 March 1674), 106 (1 June 1674), 116 (7 August 1674), 120 (8 September 1674), 129 (6 November 1674), 257 (17 November 1676), 262 (13 December 1676), and 335 (18 December 1677); Corporation of London Records Office Journal 46, folio 133v (30 November 1666); Corporation of London Records Office Repertory 74, folio 75r (28 January 1669), 189r (10 June 1669), 76 folios 58r (26 January 1671) and 72v (14 February 1671); Corporation of London Records Office, City Lands Committee Orders Books I, folios 42r (3 August 1670) and 46r (9 November 1670); Corporation of London Records Office, City Lands Committee Minutes (Rough) dated 9 October 1672 and 6 November 1672; Corporation of London Records Office, City Lands Committee Papers, MSS 158 (15 November 1676), 161 (13 December 1676), and 187 (19 December 1677).
188. Corporation of London Records Office ex-Guildhall Library MSS 322/8 and 17.
189. Corporation of London Records Office ex-Guildhall Library MS 277 folios 2v–3v.
190. Hooke 1935, 76.
191. On 2 October 1673 Hooke received £25 from the City for his work on the Monument (Hooke 1935, 63). On the same day the City ordered eight guineas to be paid to Hooke for his advice on the Rebuilding Acts (Corporation of London Records Office, City Cash Account Books I/15, folio 59r) which Hooke received on 31 October 1673 (Hooke 1935, 67). On 24 December 1675 he received a warrant from the Comptroller for £100 (Hooke 1935, 203).
192. Hooke 1935, 263. When Hooke received his warrant for the payment of £50 on 6 April 1677 he was not sure if it was valid. It was not warranted by the City Lands Committee, but by the individual members of the Committee with whom Hooke and Oliver had spoken after the Committee had met on 20 December 1676.
193. Hooke 1935, 265.
194. Hooke 1935, 328 (16 November 1677).
195. Hooke 1935, 335 (22 December 1677), 390 (26 December 1678) and 400 (24 February 1679).
196. The Chamberlain was ordered on 18 December 1682 by the City Lands Committee to pay Hooke for his work for the Committee in 1682: 'You are forthwith to pay to Robert Hooke one of the Cities Surveyors the sume of five pounds in full for his service and attendance upon and by Order of the Committee for the Cities Lands this last yeare' (Corporation of London Records Office Misc. MS 157.15). Hooke's receipt for £5 is dated 18 June 1683 (ibid.) and the payment appears in the accounts for the period ending Michaelmas 1682 under the heading of 'Guifts and Rewards' (Corporation of London Records Office, City Cash Account Books I/17, folio 221r). Another payment of £5 was made for the year ending Michaelmas 1683 (ibid., folio 283r).
197. Amidst his list of creditors Hooke writes 'I owe Mr. Loach for velvet coat and lyning £5' (Hooke 1935, 265, dated 31 December 1676).
198. Guildhall Library MS 25 548, 17–19.
199. Batten 1936–7.
200. Stoesser-Johnston 1997.
201. Jeffery 1996.
202. As Anthony Geraghty points out (Geraghty 2001) 'Woodroffe' is common in modern usage, but 'Woodroofe' was generally used in documents of the time.
203. Jeffery 1996, 31–41, 60–5, 175–7, and passim.
204. Stoesser-Johnston 1997, part 1, 65–8; Jeffery 1996, 204–5, 219–21, and 240–2.

205. Colvin 1995, 506–10.
206. Cooper 1999, 145–7 and 241–2.
207. Gunther 1935, 195, 197, 199, 200, and 202 inter alia.
208. For examples of these see respectively Gunther 1935 (69, 92–3, 98, 105, 107, 117, and 206), (89 and 108), (114 and 205), (118 and 253), and (113), where Hooke records receiving twenty guineas.
209. Hooke's career earnings based on present evidence are shown in the following table, rounded to the nearest £50. The estimates of salaries and of fees from the fund for churches are the most reliable, within 10%. The estimate of fees from citizens is a conservative estimate, and could well be higher. The final figure in italics is the least reliable of all; it could be as much as 50% higher or lower. Hooke's income was increased further from renting or leasing property.

Source	Type of earnings	Amount
The Royal Society	Salary	£750
Sir John Cutler	Salary	£1,500
Gresham College	Salary	£1,750
The City of London	Salary	£1,050
London's citizens	Fees	£1,700
Fund for churches	Fees	£2,800
Private commissions for architecture and building	Fees	*£2,000*
Total		£11,550

2

Hooke's Instruments

JIM BENNETT

ANYONE WHO HAS STRUGGLED to read Hooke's manuscript papers, particularly those written towards the end of his life, will readily agree with John Aubrey: 'I wish he had writt plainer, and afforded a little more paper.'[1] But despite the frustrations of his ageing penmanship, the reader who strains to decipher this cramped hand seems to encounter a very real individual—intense, focused, and committed—as the words squeezed on to all the available space speak still with vitality and passion. Robert Hooke's attitude to life is often portrayed in terms of personal disenchantment. Promoting this view of himself, he wrote in a late lecture that 'the Greatest part of Learned men Respecting the Reward; Soon list themselves into the Societys of Divines, Lawyers or physicians, where their way to Canaan is already chalkd out.' He, on the other hand, had found himself among 'Some straglers [who] chance to be left behind by the Caravan'.[2] But when Hooke took a broader view of human history, in his heart he was a passionate optimist. He had enormous confidence in the future, writing with intensity and verve on the potential of human progress, with style and humour about the achievements that awaited scientists in his future, and occasionally with generosity about their accomplishments in his present. Unlike his sometime collaborator, sometime antagonist, the first Astronomer Royal, John Flamsteed, who held out little hope that anyone would be able to continue his work,[3] Hooke had great faith in his successors. He despised any argument along the lines that something cannot be done because we do not have the means of doing it now, that we cannot expect some result in the future and may not try to imagine its consequences because we are not able to effect it today. A youthful enthusiasm was sustained to the very end and, if this is not the Hooke we have come to know, we have not been paying sufficient attention to his lifelong passion for instruments.

Instruments are found everywhere in Hooke's work, stretching across the many subjects that engaged his attention during a long career. Some, such as the microscope, the air-pump, the spring-balance watch, and the wheel barometer, are already firmly linked to Hooke's popular reputation, but what is striking as we expand a broader survey is the ubiquitous character of instrumentation in Hooke's oeuvre. This suggests that instruments, particularly if we include what might today be classed as apparatus, machines, demonstrations, gadgets, and practical devices, were not simply an adjunct to his natural philosophy but something that was central to his method and inseparable from his approach to nature.

It is not only Hooke's practical or experimental approach to nature that seems inseparable from instruments: they also inform his thinking as a theorist about the natural world and help to shape his most general notions about the history of human knowledge and its prospects. That general context—Hooke's own rationalization for his instruments—is worth attending to before we embark on a more detailed survey. It offers an explanation—Hooke's explanation—of why we have to give instruments such an important position in a study of Hooke, significantly more than any other scientist of his time, and perhaps even any before him.

This explanation for the prominence of instrumentation does not have to be teased out of his writings and assembled through a process of examination, analysis, argument, and comparison: Hooke's own account is pretty well sorted and systematized by the time he writes the preface to his *Micrographia*. There he begins with the human condition, the chief characteristic of which is that it is fallen: 'every man, both from a deriv'd corruption, innate and born with him, and from his breeding and converse with men, is very subject to slip into all sorts of errors.'[4] The 'deriv'd corruption' is the result of the fall of Adam, without which we would have no need of microscopes and telescopes, since everything in nature would be open to our senses and transparent to our reason. 'Breeding and converse' reflects a different source of error and shows the influence of Francis Bacon, who held that we are all too prone to accept any manner of foolish notion under the influence of common opinion, of intellectual authority, or of fashion.

Instruments are the answer to 'deriv'd corruption': 'By the addition of such artificial Instruments and methods, there may be, in some manner, a reparation made for the mischiefs, and imperfection, mankind has drawn upon it self.'[5] Instruments then are for rectifying our decayed and corrupted senses, 'supplying of their infirmities with Instruments, and, as it were, the adding of artificial Organs to the natural'.[6] Hooke's optimistic vision of the new world revealed by instruments and of the potential they unlock for natural philosophy breaks through immediately, as they carry us swiftly from depravity and corruption to knowledge and progress. The improvement of one of the corrupted senses, that of sight,

> has been of late years accomplisht with prodigious benefit to all sorts of useful knowledge, by the invention of Optical Glasses. By the means of Telescopes, there is nothing so far distant but may be represented to our view; and by the help of Microscopes, there is nothing so small, as to escape our inquiry; hence there is a new visible World discovered to the understanding. By this means the Heavens are open'd, and a vast number of new Stars, and new Motions, and new Productions appear in them, to which all the antient Astronomers were utterly Strangers. By this the Earth it self, which lyes so neer us, under our feet, shews quite a new thing to us, and in every little particle of its matter, we now behold almost as great a variety of Creatures, as we were able before to reckon up in the whole Universe it self.[7]

Recovering and enhancing the senses seems an appropriate way to construe the role of optical instruments, especially microscopes and telescopes, but Hooke intends it to apply to all artificial aids to perception, including devices which unfold the nature of mechanical operations, instruments more for accurate measurement than for observation, accessories that magnify readings mechanically, such as forms of micrometer or the wheel barometer, or even instruments that detect and measure qualities that cannot be perceived directly by the senses, such as the barometer itself or the magnetometer. Elsewhere Hooke neatly incorporates a range of meaning when he refers to the errors committed by the 'naked eye and unmachined hand'.[8] When he took charge of the Journal Book of the Royal Society, on becoming Secretary after the death of Oldenburg in 1677, he put on record there a similar explanation of his frequent use of instruments, saying that an instrument 'inlarges the empire of the senses'.[9] Instruments are for recovering, at least in part, the link with nature that was broken in Eden, by restoring the performance of that fallen and degenerate creature, man.

After dealing with improvements in sight, in the preface to *Micrographia*, Hooke runs through the other four senses, claiming that they too can be enhanced by artificial organs. Ear trumpets might be improved, of course, but Hooke points out also that sound can be carried 'through other bodies then the Air', citing experiments he has made with long stretched wires, which could be bent through many angles and still convey sound more quickly than in the air.[10] He has experimented with improving the sense of smell, noting that other creatures are able to detect specific properties of bodies, such as whether they are wholesome or poisonous, but he extends this line of thought to techniques for artificial smelling—smelling by machine. What is smelling but the detection of effluvia? Changes in the constitution of the air are reflected in the small variations in the height of a mercury column, which Hooke can magnify by his wheel barometer. Another means of 'discovering the effluvia of the Earth mixt with the Air' is by his hygroscope, detecting changes in the moisture content, 'which the Nose it self is not able to find'.[11]

Taste, like smell, is recast in a very general formulation:

> the business of this sense being to discover the presence of dissolved Bodies in Liquors put on the Tongue, or in general to discover that a fluid body has some solid body dissolv'd in it, and what they are; whatever contrivance makes this discovery improves this sense.[12]

A great deal of chemistry can thus be included in Hooke's programme for recovering the sense of taste, since where it is concerned with the contents of solutions, 'what is that discovery but a kind of secundary tasting.' Touch is 'a sense that judges of the more gross and robust motions of the Particles of Bodies',[13] so whatever detects the minute mechanical arrangement of matter, such as a microscope, improves this sense. Since heat and cold are derived, in Hooke's mechanical view of the natural world, from the motions of particles, thermometers are artificial organs of touch.

Having dealt with the senses, Hooke moves on to improving memory and reason. But the beneficial role of instruments in natural philosophy is not confined to the concept of artificial organs. In the sense that includes mechanical devices and demonstrations, instruments are the answer also to the baneful influence of 'breeding and converse':

> all the uncertainty, and mistakes of humane actions, proceed either from the narrowness and wandering of our Senses, from the slipperiness or delusion of our Memory, from the confinement or Rashness of our Understanding . . . These being the dangers in the process of humane Reason, the remedies of them all can only proceed from the real, the mechanical, the experimental Philosophy.[14]

This philosophy he contrasts with 'the Philosophy of discourse and disputation'. 'The truth is,' he exclaims, 'the Science of Nature has been already too long made only a work of the Brain and the Fancy: It is now high time that it should return to the plainness and soundness of Observations on material and obvious things.'[15] So we must turn away from converse; manipulative experience of the plain material world is to characterize the science or understanding of nature as well as the observation of nature. Instruments and mechanical devices, a familiarity with their working and an appreciation of their actions, will inform an understanding of nature that is real, mechanical, and experimental. It is interesting to note that Hooke sees this as a *return* to plainness and soundness, a restoration of something we have lost, a return to an open and rational understanding made possible by the restoration of a plain and evident observation of the world through our artificially enhanced senses.

The very notion of an instrument of natural philosophy, such as an air-pump, depends on nature being mechanical, and therefore the true natural philosophy being a mechanical discourse. Manipulating the natural world by a machine will reveal nothing of its inner workings unless it too is a machine; otherwise the natural and the artificial will simply fail to engage. For Hooke's mechanical philosophy they are fully commensurate, and the difference is only a matter of scale. Thus the instrument reveals the inner workings of nature, the natural on the micro scale, by greatly enhancing our senses, and what it reveals is that the micro-world is mechanical, governed by the same principles as the macro, and so to be understood and explained in terms of the gross machines of our macro-world. In a passage from the same part of the preface to *Micrographia*, Hooke makes this connection absolutely explicit:

> It seems not improbable, but that by these helps [optical instruments] the subtilty of the composition of Bodies, the structure of their parts, the various texture of their matter, the instruments and manner of their inward motions, and all the other possible appearances of things, may come to be more fully discovered . . . From whence there may arise many admirable advantages, towards the increase of the Operative, and the Mechanick Knowledge, to which this Age seems so much inclined, because we may perhaps be inabled to discern all the secret workings of Nature, almost in the same manner as we do those that are the productions of Art, and are manag'd by Wheels, and Engines, and Springs, that were devised by humane Wit.[16]

Thus Hooke sees a continuity that runs through natural philosophy and mechanics, through the natural and the artificial, and the common character is not only mechanical but also operative. Doing is closely entangled with knowing. For Hooke, the intelligibility of nature and the explanations to be devised by the natural philosopher depend on instruments in this broader sense, just as the enhanced observation and more acute examination of nature is instrumental. We shall see in the chapter that follows this one how Hooke's natural philosophy was fundamentally mechanical, that phenomena were to be explained in terms of matter and motion, and hypothetical accounts of the appearances of things were to be formulated in terms of invisible micro-mechanical action. But Hooke was unusual in the extent to which he was not content to rest with verbal formulations of mechanical hypotheses; for him a working demonstration was an integral part of rendering an explanation intelligible, cogent, and convincing. Instruments and mechanical devices were much more than illustrative adjuncts to Hooke's programme for natural philosophy: they were also inseparable from his practice, as we shall see from a review of their occurrence throughout his work.

Mechanical Horology

Hooke's reference above to wheels and springs devised by human wit brings to mind watches, one of the most contentious and contested aspects of Hooke's interest in instruments and mechanics. Partly because of the secrecy adopted by Hooke and others, partly on account of the passions and jealousies the subject aroused, conclusions are still difficult to reach and far from certain or complete, but our situation is much improved by Michael Wright's systematic and careful assessment of Hooke's contributions.[17] Wright focuses only on Hooke's attempts to design a longitude watch, but this brings in some of the central issues of the design of escapement and of maintaining power, and especially the application of a balance spring.

Hooke had certainly devised forms of balance spring before he heard of Huygens's design in 1675. Wright arrays and assesses the many references to Hooke's work on spring-regulated watches in the 1660s, including his disclosure of at least part of his design to William Brouncker, Robert Moray, and John Wilkins by January 1664. One of the more revealing references is a record of Lorenzo Magalotti's visit to the Royal Society in February 1668, where he was shown a watch, 'the time being regulated by a little spring of tempered wire which at one end is attached to the balance-wheel, and the other to the body of the watch'.[18] A source of confusion for the modern reader is that this device is referred to as 'a pocket watch with a new pendulum invention', but this invokes the pendulum in a conceptual or analogical sense, not a material one. Hooke also spoke of supplying his watch with a form of 'artificial gravity', thinking again in terms of the swinging action of a pendulum under the restoring or returning influence of gravity.

Christiaan Huygens had applied the pendulum to the common arrangement of the 'verge' escapement—the device that restrains the motion of a watch or clock as the force of the weight or driving spring is released through the train of wheels, the wheels themselves being arranged so as to display the time by attached index hands. The escapement is designed to push a balance arm or wheel first one way, then the other, but without some restoring force applied to the balance itself, it moves to and fro at the mercy of the escapement, its inertia merely restraining the circular motion imposed on the escape wheel by the driving force transmitted through the train of wheels. Huygens's pendulum began to regulate this motion, though it could not at this stage avoid mechanical interference from the escapement and so impose its natural oscillation completely on the timekeeping of the watch. Hooke's spring was a different way of providing the balance with a restoring force once it was displaced and of giving it a natural oscillation of its own; he conceived this as an 'artificial gravity', by analogy with the pendulum formerly used by Huygens. Hooke had a variety of shapes and arrangements of springs, acting singly or in pairs, and although they are not clearly recorded, he does not seem to have anticipated Huygens's volute form, which became the common design.

The advantage of this 'artificial gravity' was that it made the timepiece much more portable, the pendulum being easily deranged by the motion of a ship, but Hooke included other horological novelties in his longitude watch. To deal with the derangement of the horizontal oscillator in his watch, equivalent to the disturbance of a pendulum, namely by a rapid rotation of the frame, he introduced two balances whose motions are connected and exactly out of phase. He described the first known constant-force escapement, with the aim of avoiding the variations in the force transmitted to the escapement by the train. He used this force to set a mechanism—Hooke likened it to the spring-loaded cock of a gun—which, when released in the oscillatory cycle, acted independently and directly on the palets to give impulse to the balance. The precise form of Hooke's device is not known, but the concept was clearly stated and did become important in subsequent horology, with many later designs coming into use.

Hooke also produced the first design for a 'maintaining power' for use with a fusee. The fusee is a conical pulley with a spiral groove to receive a line coming off the cylindrical surface of the barrel containing the mainspring. The arrangement is such that the fully-wound spring delivers its force, via the line, closest to the axis or arbor of the fusee, and progressively further away as it unwinds and its force decreases. The result, following the law of the lever, is that the torque or turning force of the arbor and the great wheel at one end of the train is fairly constant through the unwinding of the spring. This arrangement had been used since the first half of the sixteenth century, and Hooke's device, the maintaining power, was a modification to prevent the watch stopping as it was being wound. In the usual arrangement, the fusee arbor is connected to the great wheel via a ratchet, which allows the watch to be wound by turning this arbor in the direction opposite to its normal motion, so as to take the line from the barrel back on to the fusee, turning the barrel in the process and coiling the spring inside. As this happens, no force is given to the train and the watch will stop. Hooke's concern that 'there shall be noe time lost in the winding of it vp'[19] led him to have the great wheel mounted directly on the fusee arbor and the barrel carried around the fusee as the watch is wound, the line transferred, and the spring coiled.

On 18 February 1675, the very day that Isaac Newton was admitted to fellowship of the Royal Society, Hooke notes in his diary, 'Zulichems [that is, Huygens's] spring watch spoken of by his letter. I shewd when it was printed in Dr. Spratts book. The Society inclind to favour Zulichems.'[20] Hooke also notes that 'Mr. Newton told me his way of polishing metall on pitch,' which we will come back to later.

Now Hooke began a campaign to establish his priority, showing his designs to his friends, seeing Thomas Tompion very often and arranging experiments and watch-making with him, 'Discoursd much with Sir Ch. Wren about spring watch,'[21] and so on. Only two days after the Royal Society meeting, he found that the register confirmed that he had demonstrated a spring-regulated watch in 1666.[22] He was gratified to discover in the record of the meeting on 29 August that 'Mr. Hooke produced also a new piece of watch-work of his contrivance, serving to measure time exactly both by sea and land.'[23] He began to doubt the good faith of Henry Oldenburg, and on 3 April heard that Oldenburg was involved in a patent application for Huygens's watches.[24] Four days later he was presenting Charles II with a spring watch made for him by Tompion: 'The King most graciously pleasd with it and commended it far beyond Zulichems. He promised me a patent.'[25] Although Hooke dined with Brouncker that day, on the following day, a Royal Society meeting day, he discovered that Brouncker was in on the patent application with Oldenburg and he faced up to them at the Bear tavern: 'I vented some of mind against Lord Brounker & Oldenburg. Told them of Defrauding.'[26] In the year of Oldenburg's death, Hooke repeated the accusation against him in print in his tract *Lampas* of 1677, citing an intention to defraud by 'one that made a trade of Intelligence'.[27] The episode is cast as a contest between honour and duplicity, between 'the Describer of Helioscopes' (Hooke) and 'the Transcriber of Intelligence'.

The argument over the spring-regulated watch is well known and we need not pursue it further. In fact Hooke had another source for a restoring force, by having a magnetic balance bar moving under the influence of a loadstone. Horologists have paid little attention to this, perhaps because it seems such a bizarre idea, but it does show the value of the breadth of Hooke's activities in natural philosophy and his facility for moving among disparate topics and carrying ideas and techniques between them, in this case between the natural philosophy of the loadstone and the mechanical art of horology. Hooke showed different versions of the design to the Royal Society from April 1666 on, so this overlapped with the period when he was developing his spring-regulated watch.[28] At least one watch was made, with Hooke devising modifications until 1669, when he would explain the latest improvement only to the President (Brouncker), 'being not yet free to declare it in public,'[29] and when it became an object to be shown to another Italian visitor, in this case the Venetian ambassador.[30] The story of this initiative is not unlike that of the spring-regulated watch, except that controversy did not make it widely known. With magnetic artificial gravity, timekeeping could be adjusted merely by altering the distance of the loadstone from the magnetic balance arm.

Hooke had many other horological designs, relating to clocks as well as watches, but many of them are recorded only in sketchy references to schemes or ideas, or sometimes working timepieces brought into Royal Society meetings. These can lead to hopeful but insecure interpretations; the former attribution of the common anchor escapement of longcase clocks to Hooke is now out of favour. But some ideas are worth noting, even where the details are slight. For example, in May 1669 Hooke showed the Society 'a new kind of pendulum of his own invention'.[31] It was 14 feet long, had a period of two seconds, a very heavy bob, moved with an amplitude of only half an inch by a tiny force, and needed winding only every fourteen months. By October he was saying that his clock, on account of its accuracy, was intended for astronomical observations,[32] while in a lecture of 1687 he referred to his experiments with this clock, eighteen years before, saying they were for detecting any irregularities in the rotation of the earth and linking them with his attempts at the same time to detect the earth's annual motion through the measurement of stellar parallax,[33] which we will consider later. Hooke's long-pendulum clock has many similarities to two installed at the foundation of the Royal Observatory in Greenwich and made in 1675 by Hooke's collaborator in horological work, Thomas Tompion,[34] while the octagonal Great Room of the building was designed to accommodate them, either by Hooke himself or by his friend Christopher Wren. Flamsteed used them to investigate precisely Hooke's question, the irregularities of the earth's diurnal motion.

At a more mechanical level, Hooke sought to improve the engagement of the wheels in the train, so that they worked evenly and regularly,[35] and also on a wheel-cutting engine.[36] Details of both of these are sketchy, but we do know more about a radically new design for a pendulum clock, dating from the period that was so productive for watches, the mid-1660s, one with a conical pendulum, that is, where the bob moves continuously in a horizontal circle instead of to and fro in a vertical arc.[37] Hooke also managed to make its motion isochronous by confining the bob to a parabola of revolution by hanging the suspension cord over a curved surface that altered its effective length according to the amplitude of the bob (Fig. 16).[38] He produced a

Fig. 16 Hooke's demonstration of the isochronous motion of a conical pendulum moving with different amplitudes. By altering the effective points of suspension, the bob is confined to a parabola of revolution. (Royal Society Classified Papers XX, no. 53.)

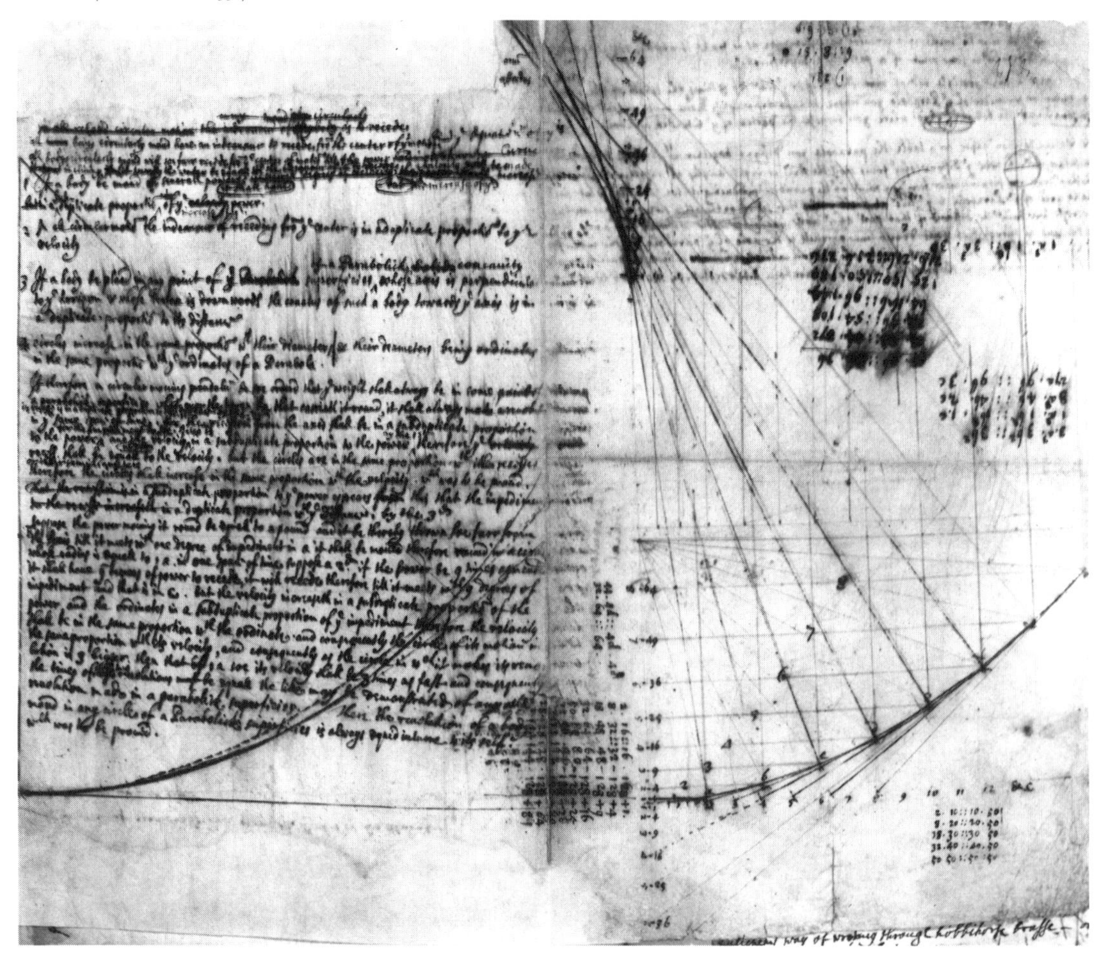

demonstration of the isochronous parabolic motion alongside a working mechanism.[39] One reason for his interest in the conical pendulum was to illustrate the combination of a rectilinear motion and a central force, and to apply this to analysing the motions of planets and comets.[40] We will come across the conical pendulum later, in another instance of transference of thought between disciplines, but for now we can note it as a radically innovative horological design. It too was later published by Huygens and, while this did not occasion so much anxiety as the spring watch, Hooke did make his point in a Cutlerian Lecture that Huygens had announced his results

> without naming me at all, as concern'd therein, though I invented it, and brought it into use in the year 1665, and in the year 1666, I communicated it to the Royal Society, at their publick Meetings, both as to the Theory and Practick thereof, and did more particularly explain the Isocrone motion of the Ball of a Pendulum, in a parabolical Superficies, and the Geometrical and Mechanical way of making the same move in such a Superficies, by the help of a Paraboloeid, which I caused also to be made and shew'd before the same Society, upon several days of their publick Meeting, where besides many of the Society, were divers strangers of forreign parts.[41]

Marine Science

One reason for the excitement generated by the watch was, of course, its relevance to finding longitude at sea. When Hooke first revealed his proposal, he did so not to the full meeting of the Royal Society but, as we have seen, to three senior figures. Brouncker made a guarded report to the Council of an unidentified invention by Hooke 'which might prove very beneficial to England, and to the world'.[42] Up to ten pounds was allowed for its development. When it all came to light some sixteen years later, in 1675, Sir James Shaen assured Hooke that a longitude solution would be worth £1,000 or £150 per annum; Hooke plumped for the latter.[43]

It is worth noting that on the day Hooke revealed his 'new piece of watch-work' to the Society in August 1666, the account he was relieved to find in the Journal Book records as the immediately following item, 'He mentioned again a perspective, which he was preparing for observing the positions and distances of fixed stars from the moon by reflexion.'[44] The minutes do not mention this, but the connection between these two, and the reason why Hooke followed one with the other, was the longitude question.

This was the instrument that Hooke, or perhaps Waller, called 'an Instrument for taking Angles at one Prospect', that is, when looking in only one direction (Fig. 17).[45] This is done by mounting a mirror on one arm of a sector and a telescopic sight on the other. One object is viewed by reflection in the mirror, the other directly by looking past the mirror. With both in the field of view, their images are brought into coincidence in the eyepiece, when the angle between the arms of the sector will be half that between the objects. Thus the degree scale on the straight rule is graduated at double the rate that might be expected from the angles between the arms. This, then, is the principle of measurement used for centuries in the most ubiquitous marine instruments for taking angles, namely octants, sextants, and reflecting circles, and Hooke's is the first published account. It fits into his programme as the seagoing tool for measuring lunar distances for longitude—one of the three components essential to this method. He was actively interested in both of the other two, establishing the positions of the stars and the theory of the lunar motion.

Hooke had other schemes for the longitude and other improvements in navigational instruments, such as a new form of portable quadrant for latitude, improvements to the backstaff, and so on.[46] Most cannot be identified further than the slight minutes of the Royal Society, but one specific idea he did cite as his own was the use of a lens in place of the shadow vane on a backstaff. The backstaff is used for measuring the altitude of the sun for finding latitude and, as part of the procedure, a vane set on a scale of degrees is used to cast a shadow on to a target. The seaman has his back to the sun, hence the name of the instrument. But in hazy weather the shadow will be indistinct and a lens focusing a spot of light on the target will give a better result:

> The Instrument which I shew'd the Society, some Years before the Sickness, by making use of a Telescope-glass, instead of the small hole or slit of the Shadow-vane of a Back-staff, but was not made use of 'till about ten Years after, and yet now it meets with general approbation, and is of continual use, and pretended to be the invention of another, tho' my shewing thereof was Printed in the History of the Royal Society.[47]

Hooke does not identify the plagiarist, but this accessory to the backstaff is known today as a 'Flamsteed glass'. 'The first discoverer,' he lamented, 'is dismist with Contempt and Impoverishment.'

Celestial navigation—finding latitude and longitude by measurements of the heavens—was at the glamorous end of contemporary interest in managing ships at sea, but Hooke also worked on the more mundane

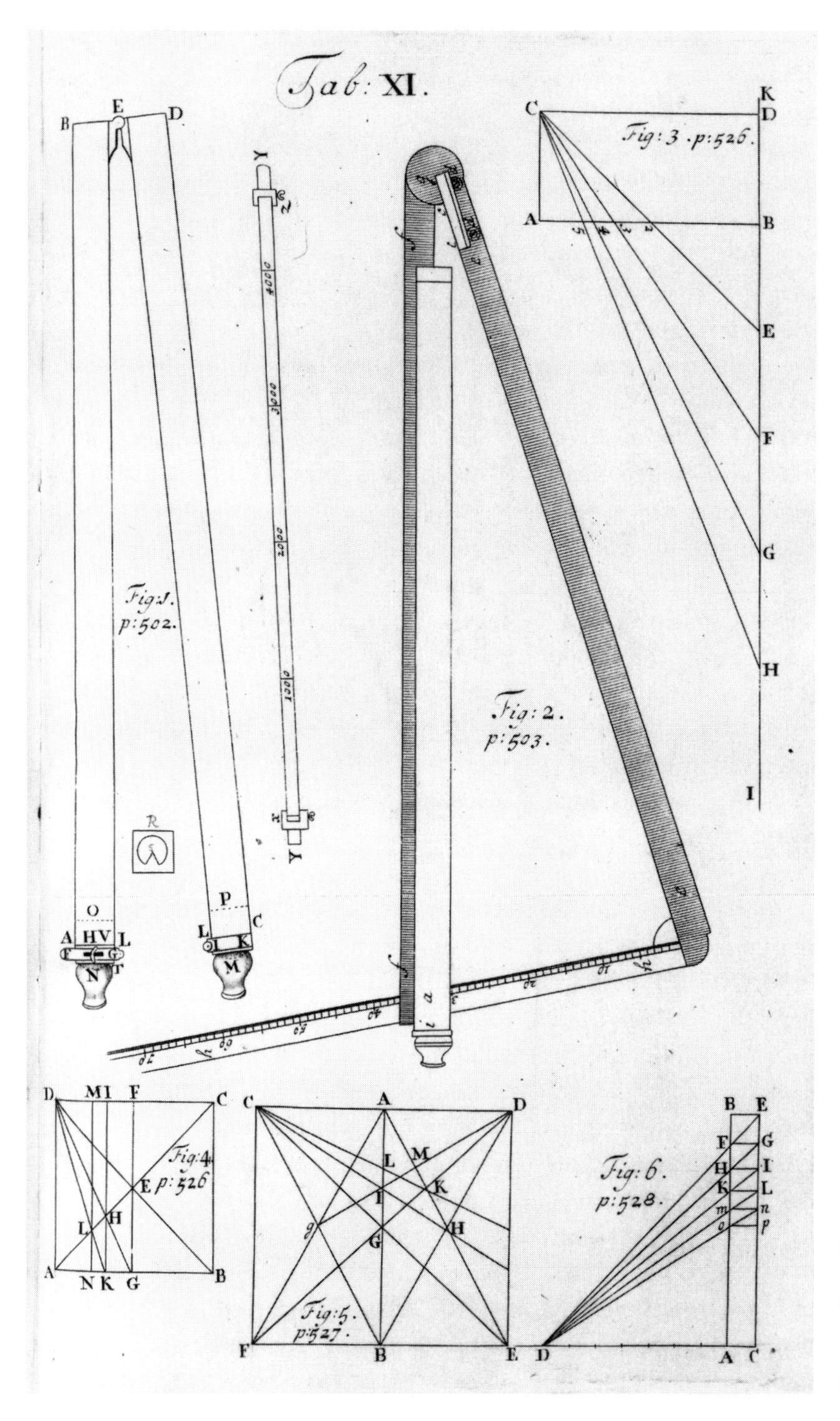

Fig. 17 The reflecting instrument (Fig. 2), using the principle later adopted for marine sextants and octants, from Hooke's *Posthumous Works*, edited by Richard Waller (1705). (Museum of the History of Science, Oxford.)

business of depth sounding, though typically his sounder was to be different: this was sounding, as he put it, 'without a line'.[48] A weight would carry a float to the bottom, attached by a wire spring, which is released on impact, allowing the float to return to the surface (Fig. 18).[49] The depth could be deduced from the time elapsed, once standard measures had been taken in known depths. Hooke claimed that the originality of his device was his arrangement for ensuring that the weight and float descended together and were reliably separated at the correct moment; the general idea was not new, but his mechanism was. Trials were made and Hooke reported their success, but the minutes solemnly record that 'Oozy ground was observed to be most likely to make them unsuccessful.'[50] To a later objection that Galileo's law of free fall indicated that time of descent and ascent would not be proportional to depth, Hooke said that terminal velocity would be reached after about two fathoms.[51]

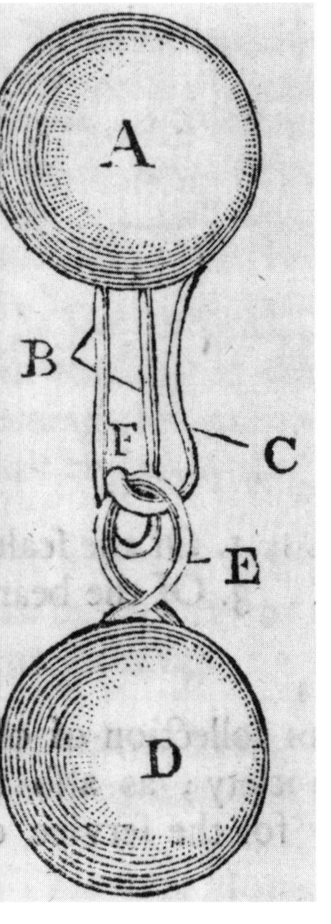

Fig. 18 Hooke's design for a depth sounder, in which a float is released as the device hits the seabed, from Thomas Birch, *The History of the Royal Society of London* (1756–7), i, 307. (Museum of the History of Science, Oxford.)

The other thing that is original about this area of Hooke's work—and here it becomes an instance of the integration in his programme of work in natural philosophy with practical mathematics and mechanics—is that he uses his basic sounder idea as a means of tackling other problems. His detachable float becomes a vehicle for bringing back other information about the deep sea in addition to its depth. It is adapted to carry specially designed mechanisms for sampling the sea at depth and the sea bed (Fig. 19).[52] According to Anita McConnell, Hooke's drawings of 1663 of a sounder and a sampler are the earliest known.[53]

Problems over the difficulty of ensuring that descent and ascent have the same velocity led Hooke to introduce radical new designs. In 1678 his approach was to make use of the variation of water pressure with depth by sinking glass tubes and measuring the length of water forced in against the air and retained as the tube was recovered.[54] This had the advantage that the result was independent of the route taken to reach the bottom or to return, and the same principle was applied successfully in sounders during the late nineteenth century, notably the one designed by Lord Kelvin. It could also be used as an instrument for measuring pressures in the sea. As late as 1691, Hooke had a new design in which his sinker carried a hollow cone to the bottom and water forced in at depth was weighed when the float was recovered.[55] In another of his late sounders, a vane rotates in passing through the water as the sounder descends and the revolutions are counted, but the vane is clamped by a lid activated as it ascends.[56] Again this principle was successfully exploited in nineteenth-century sounders. In one impossibly ambitious device he combined two of the vane or waywiser sounders, one for each direction of travel, and two samplers.[57]

This idea of combining different instruments in a single recording station is evident also in Hooke's 'way wiser for the sea'. This was a form of mechanical log with the vane and counter now trailed in the water instead of descending through it, which kept account as well of all the compass bearings. It was a complete dead reckoner for seamen,[58] though Hooke states very clearly that the problem for navigation is that it takes no account of currents: 'yet 'tis defective for finding the true way of the Ship over the parts of the Earth subjacent to that Sea, because it distinguisheth not the current or setting of that part of the Sea.'

The other outstanding example of a compendium of instruments forming a single recording station is the weather clock. The parallels are significant, for in many ways Hooke was treating the sea as he did the atmosphere, and he had instruments for measuring submarine temperature and pressure.[59] The sea and the air were media or elements requiring instrumental 'histories' in early Royal Society terminology, through the provision of 'artificial organs'.

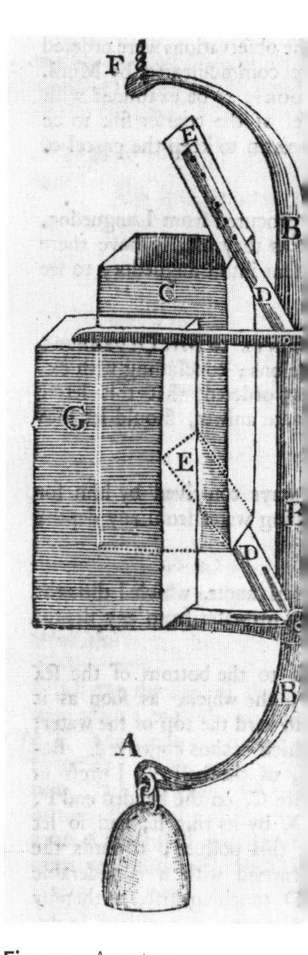

Fig. 19 A water sampler for recovering samples from the depths of the sea, from Thomas Birch, *The History of the Royal Society of London* (1756–7), i, 308. (Museum of the History of Science, Oxford.)

Meteorology

One of Hooke's inventions that links meteorology with the sea was his marine barometer. This utilized his ingenious arrangement of the siphon barometer, one of his most commonly encountered devices, since it is used also in the common wheel barometer.[60] Instead of submerging the open end of a mercury-filled tube in a reservoir of mercury, as in the classic experiment of Torricelli, the tube is bent into a J shape, the shorter limb remaining open to the air, and the difference between the two mercury levels was the measure of pressure. The wheel barometer was one of Hooke's earliest instruments, being shown to the Royal Society in 1663, and it seems likely that it arose out of work done with Boyle on the relationship between volume and pressure in air, which used a J tube in a rather different way, the shorter limb being closed to contain the air being compressed by the mercury column. It is typical of Hooke's ability to extend and modify his inventive thinking that he adapts the technical repertoire of one experiment to create a different instrument for a separate purpose. In this case he was moving, once again typically for him, between an experiment in natural philosophy and a practical instrument.

In the case of the marine barometer the longer limb of his siphon barometer has air trapped above the mercury—in the seventeenth century this was called a 'weather-glass'—and an associated sealed spirit thermometer is used to disaggregate the temperature component from the weather-glass readings to yield the pressure.[61] This principle was used in the 'sympiesometer' patented in 1818 by Alexander Aidie and marketed by him as, indeed, a marine barometer.[62]

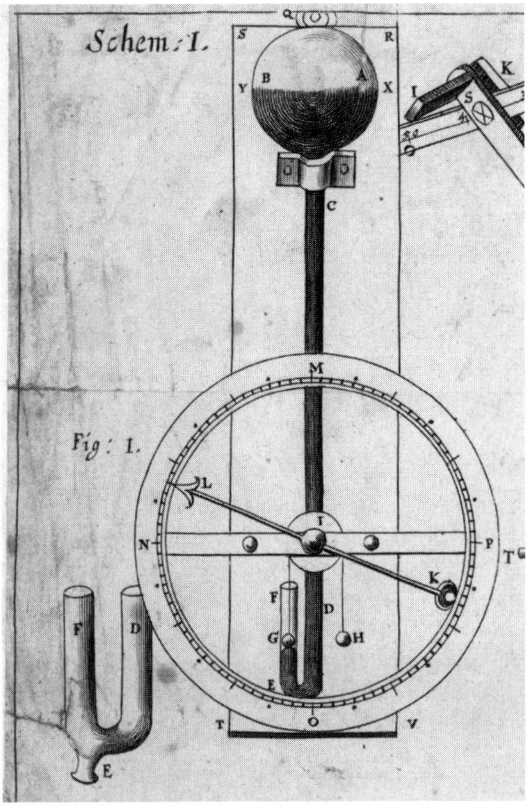

The wheel barometer, one of several ways Hooke proposed for magnifying the small movements in the height of a mercury column, is the best known of Hooke's meteorological instruments, since it was widely used in a commercial design for the domestic barometer (Fig. 20).[63] This again uses the siphon barometer arrangement and Hooke has a cistern at the top of the closed limb so that the difference in levels is mostly reflected by the mercury in the smaller tube. Here a float resting on the surface of the mercury has an attached thread that runs over a pulley wheel to a counterweight, while fixed to the axis of the wheel is an index moving over a circular scale. The diameter of the pulley can be chosen to magnify the original motion of the float to a degree appropriate to the index and scale.

Fig. 20 One of Hooke's designs for a wheel barometer, as published in *Micrographia* in 1665. (Museum of the History of Science, Oxford.)

Hooke had other ways to magnify barometer readings. He introduced a liquid above the mercury in the open tube, but constricted its motion into a narrower tube so as to oblige it to cover a greater range. He tries two immiscible liquids, the upper extending to a second cistern but meeting the lower in a constricted tube, where the reading was taken.[64] The rationale behind this concern with magnification derived from an exchange between two of Hooke's close associates, Boyle and Wren, who discovered the small variation in the height of a mercury column through wondering whether it might be possible to detect an increased pressure prompted by the passage of the moon and postulated by Descartes as the cause of the tides. There was indeed a variation, but it had nothing to do with the moon, and it was thought by Wren and others that this might be a clue to some general state of the atmosphere, also related to health and epidemical disease. Until then the main interest in the Torricellian tube or mercury column had been its overall height; now it became important to focus on the small variations in a stationary tube, which Hooke magnified with his wheel barometer. In fact, Wren had already done the same for the weather-glass.

Other individual meteorological instruments Hooke worked on included sealed thermometers, bringing them, as he put it, to 'a great certainty and tenderness'.[65] Here he was following on the work of the Accademia del Cimento in Florence, but Hooke insisted on the importance of what he called his 'thermometrical standard'; it is not clear whether there was any standardization in the Florentine instruments. A range of his individual instruments was brought together in his weather clock, a design to combine his automatic rain gauge,[66] wind gauge,[67] hygroscope, wheel barometer, thermometer, sunshine recorder, and anemometer so as to produce an automatic recording station driven by a pendulum clock and generating a paper record.[68] Many of these were actually included in his working machine of 1679. Here again Hooke was extending the work of Christopher Wren.

Hooke had several forms of hygrometer, but he was particularly intrigued by the hygroscopic properties of the beard of a wild oat and he used them to make a very sensitive instrument (Fig. 21).[69] He gives a microscopical account of the beard in *Micrographia*, before moving to a description of his instrument with evident pleasure and fascination at making so sensitive a device from such a commonplace material.

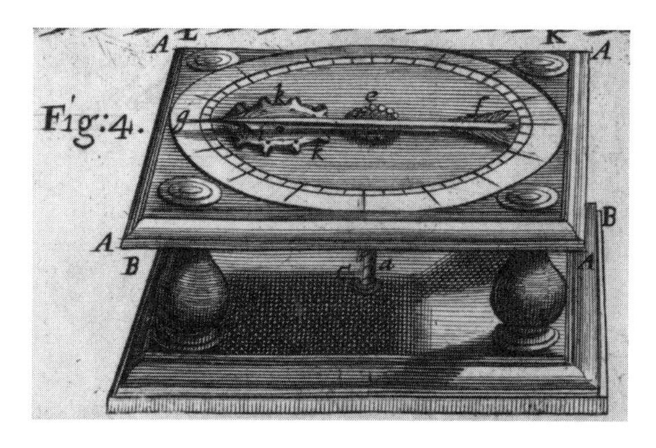

Fig. 21 The hygrometer as it appeared in *Micrographia*. Both the wheel barometer (Fig. 20) and the hygrometer also appeared, in slightly different forms, in Sprat's *History of the Royal Society* (1667). (Museum of the History of Science, Oxford.)

> I have made several trials and Instruments for discovering the driness and moisture of the Air with this little wreath'd body, and find it to vary exceeding sensibly with the least change in the constitution of the Air.[70]

Breathing on it was enough to turn the index completely around the dial. Hooke's enthusiasm must have been infectious, for he was instructed to include his hygroscope in the entertainment the Royal Society was planning for the King in 1663.[71] Several times Hooke brings his reader back to his delight in the instrument's sensitivity and eventually admits that he suspects some fundamental discovery lies behind this phenomenon.

> This, had I time, I should enlarge much more upon; for it seems to me to be the very first footstep of Sensation, and Animate motion, the most plain, simple, and obvious contrivance that Nature has made use of to produce a motion, next to that of Rarefaction and Condensation by heat and cold. And were this Principle very well examin'd, I am very apt to think, it would afford us a very great help to find out the Mechanism of the Muscles.[72]

Again he shares his fascination with the responsiveness of his instrument: 'there is nothing else requisite to make it wreath and unwreath it self, and to streighten and bend its knee, then onely a little breath of moist or dry Air, or a small atome almost of water or liquor.' With the beard held near to a fire, the tip of a tiny shred of paper dipped in alcohol will make it untwist, then retwist as the alcohol evaporates, untwist as it is touched again, as often as you wish: 'so may, perhaps, the shrinking and relaxing of the muscles be by the influx and evaporation of some kind of liquor or juice.'[73] It is wholly typical for Hooke to move from a microscopical description to an instrument—in this case to one, as we have seen, for artificial smelling—and from there to an ambitious projection of the greatest generality. Just as typical is the postponement of a full account: 'But of this Enquiry I shall add more elsewhere.'

Natural Philosophy and Mechanics

The meteorological instruments for the sea were practical devices of use to mariners at the mercy of wind and storm, but those on land were examples of that radical notion, the instrument of natural philosophy. Such an instrument brought with it philosophical and methodological baggage. It implied an experimental methodology and, depending on the instrument, might also entail a mechanical attitude to nature—Hooke's 'real, experimental, mechanical philosophy' (see above). Traditionally instruments belonged in mathematical science—they were astrolabes, sundials, theodolites, and so on—and they had no business in natural philosophy. Hooke, of course, was not the first to use instruments in his natural philosophy, but the extent of his commitment, the comprehensive integration of these two disciplines, reaches its full and enthusiastic expression in Hooke.

The air-pump, or 'pneumatical engine' (see Fig. 51), designed for Boyle by Hooke, as a thorough remaking of Boyle's previous pump, is one of the earliest instances we have of an instrument by Hooke, and at the same time the clearest example of his role as, it might be said, a natural mechanic. He dated the work to 1658–9 and Boyle published an account of it in 1660.[74] An essay in glass, brass, and wood, together with miscellaneous substances used as sealants and lubricants, it was a commission unlike any other in contemporary England. It fitted into no known category of manufactured object, but demanded familiarity with the capacities of different trades. It required an appreciation of the needs of the client as natural philosopher, combined with an ability to communicate effectively with artisans in the precise manufacture of components. This latter skill is not trivial and there are many contemporary instances where its lack thwarted the production of an instrument; most often these are ascribed to the inability of the workmen, on the word of the only participant in the failed transaction to have recorded a judgement. Hooke encountered unsatisfactory workmen, of course, but he also knew how to recognize, value, cultivate, and work alongside a skilled artisan. In the case of the air-pump he had to go to London for the brass cylinder and other specific parts; this was probably his introduction to such negotiations, in which his aptitude would later be vital to the experimental work of the Royal Society. It is difficult to imagine that anyone in England other than Hooke could have carried Boyle's commission to a successful outcome.

Hooke also designed and used an instrument working in reverse, his 'engine for the condensation of the air' for experiments on the effects of elevated pressures.[75] He had improved hydrometers,[76] refractrometers for liquids and for solids,[77] a model of the eye,[78] a model of muscular action,[79] and so on. It is interesting that his mechanical muscle acted by the application of heat and cold, which we saw him mention as the only more effective cause of rarefaction and condensation in nature than the one witnessed in the hygroscopic action of the beard of wild oat, which he also likened to the action of a muscle. An early paper involving his hydrometers, dating from January 1663, is an example of the ambitious implications that could flow from Hooke's consideration of a relatively humble instrument, essentially a weighted and graduated float.[80] Here his treatment of the different densities of warm and cold water expands first to global and then to cosmic implications—the former when he deals with the loading of ships in polar and equatorial seas, the latter as he discusses the planets acting as hydrometers in the fluid vortex of the heavens.

Other instruments of natural philosophy included a new form of variation compass to measure the declination of the earth's magnetism,[81] and a paired magnetometer and gravimeter to compare the force laws of magnetism and gravity, 'it being probable, that if they hold the same proportion, they have the same cause'.[82] What that cause or those causes might be seemed for the present profoundly unknown, those who had pronounced on the question revealing nothing except our shared ignorance:

> some making it Corporeal, some Spiritual; but what either of them mean either by Corpuscles or Magnetical Effluvia, or Atoms, or Magnetick Vertue, or Hylarchick Spirit, or Anima Mundi, when you come to inquire to the bottom you find, that neither they nor we know what is meant.[83]

So the task for the time being is to suspend speculation and look to our instruments.

It was following Hooke's demonstration of his magnetometer—'apparatus for to shew by experiments the strength of the loadstone's attraction, and to find in what proportion it draws, at several distances'[84] —in December 1673, that William Petty made a clear and unambiguous proposal for the creation of what would later be thought of as a cabinet of physics, and reproduced in many institutional and private settings in the eighteenth century:

> Upon this occasion Sir William Petty moved, that the Society would give orders, that there might be a constant apparatus of instruments ready for the making of several kinds of experiments depending on several heads; for instance, for experiments of motion, optical, magnetical, electrical, mercurial, &c. And that such instruments, as had been formerly used by the Society, and were out of order, might be repaired, and all these put together in a room by themselves, to be ready upon occasion for strangers, or for repetition and farther prosecution of the several sorts of experiments.[85]

Whether this is the first scheme for a recognizable cabinet of this type is far from certain—the Accademia del Cimento, for example, had a collection of instruments—but Petty's proposal in substance and in detail is both early and accurate, and it is no accident that it arose in the setting of the Royal Society. Here experimental philosophy was to be collaborative, public, and accessible, not to be confined to the private workshop of discovery, which might resemble too closely the den of the alchemist or the studio of the magician. These experiments were to be performed before witnesses and repeated at will, so they created a need for a 'standing' potential for demonstration, that is, for a cabinet of natural philosophy. As Hooke produced instrument after instrument and showed them at the Society's meetings, the case for such a resource would have seemed compelling.

As might be expected from what we have already seen, Hooke's instruments within the traditional learned discipline of mechanics—dealing with force and motion—also have a natural philosophical agenda. There was an instrument for measuring the times of falling bodies,[86] that is, for demonstrating Galileo's laws of free fall, and one for measuring the force of falling bodies (Fig. 22).[87] The latter he immediately extended into natural philosophy, since impact was fundamental to all action in the mechanical philosophy, and the laws of impact enunciated by Descartes became a target for his instrumental results in February 1663.

If he could investigate elastic impact with his instrument, he could move his project from mechanics understood as a mathematical science into mechanics as the basis of a natural philosophy, a move he makes perfectly explicit:

> Now as exact trials of this kind may be very useful in mechanics, so could they be made with bodies perfectly solid, would they be for the establishment of one of the chiefest philosophical principles, namely, to shew the strength, which a corpuscle moved has to move another . . . [he cites Descartes's principle that a smaller moving body cannot move a larger stationary one] . . . yet these experiments do seem to hint, that the least body by an acquired celerity may be able to remove the greatest; though how much of its motion is imparted to the bigger body, and how much of it is recoiled into the smaller, be not determined by these experiments.[88]

Hooke may have known of Wren's alternative account, which contradicted Descartes: it had been demonstrated before the Society but had not been published.[89] Either way, he was carrying his mechanical instruments into the heart of contemporary natural philosophy. His 'Philosophical Scales' (Fig. 23), an application of the law associated with his name, 'Ut pondus sic tensio', were, he said, 'of great use in Experimental Philosophy'.[90] He designed them to investigate the force law of gravitation in the early 1660s, a subject which did indeed bring him to the new locus of natural philosophical interest, now in contest with the new censor of Descartes's cosmology, Isaac Newton.

In other areas of practical mathematics, Hooke devised instruments for drawing[91] and for surveying.[92] He had a waywiser for use with a carriage; like his waywiser for the sea, it recorded direction as well as distance, and in this case produced a trace on paper.[93] He designed levels, including early, though not the earliest, designs for bubble levels, but they may not have been specifically for surveying.[94] There was a host of technological devices, some of which we might admit as instruments, such as his 'engine for determining the force of gunpowder'[95] or his mechanical

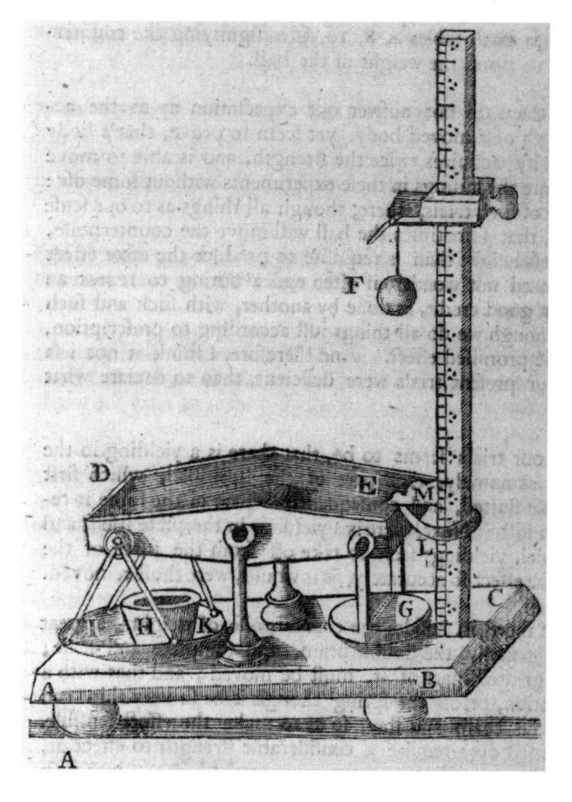

Fig. 22 The instrument for measuring the force of falling bodies, from Thomas Birch, *The History of the Royal Society of London* (1756–7), i, 194. (Museum of the History of Science, Oxford.)

calculator,[96] while others seem to belong in technology, such as his 'new cyder engine',[97] his air-gun,[98] his whale-shooting engine,[99] his windmills,[100] his turning and dividing engine,[101] and so on. It is not clear that the distinction would have meant much to Hooke, and all were reported to the Royal Society. His lens-grinding engines were relevant to telescopic optics,[102] while his series of lamps, designed to deliver oil to the flame in a steady supply as the reservoir is emptied, were useful devices, but also essays in the mechanical discipline of hydraulics, and indeed charming examples of practical design (Fig. 24).[103] Typically for Hooke, they also have a natural philosophical application, in the control of 'philosophical furnaces' for experiments requiring long periods at a constant temperature, such as for chemical reactions or incubating eggs.[104]

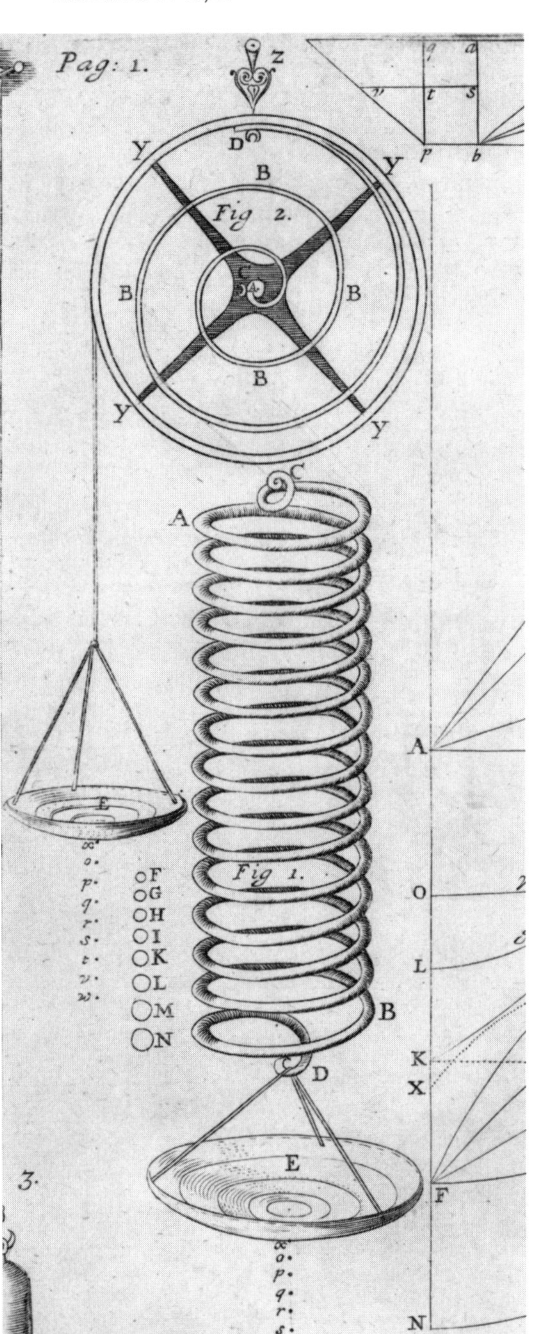

Fig. 23 Two designs for 'philosophical scales', using a coiled spring and a helical spring, published by Hooke in *De Potentia Restitutiva* of 1678.

Hooke's commitment to mechanical work, stretching as it did across mechanics in its traditional sense, practical technology, and natural philosophy, seems to bring the practical and intellectual outcomes of his industry into a single enterprise. This was evident also to first-hand observers. His friend Richard Waller, Secretary of the Royal Society, who wrote a biography soon after Hooke's death, interpreted his juvenile facility for mechanical toys in a way we can be sure Hooke would have warmly approved:

> This early Propensity of his to Mechanicks was a sign of his future Excellency in such Contrivances, and admirable Facilty he afterwards manifested in applying Mechanical Principles to the explication of the most difficult Phaenomena of Nature, and I remember it has been often observed by several Persons, that whatever apparatus he contrived for the exhibiting and Experiment before the Royal Society, it was performed with the least Embarrassment clearly and evidently, to explain the present Subject, which was a sufficient proof of his

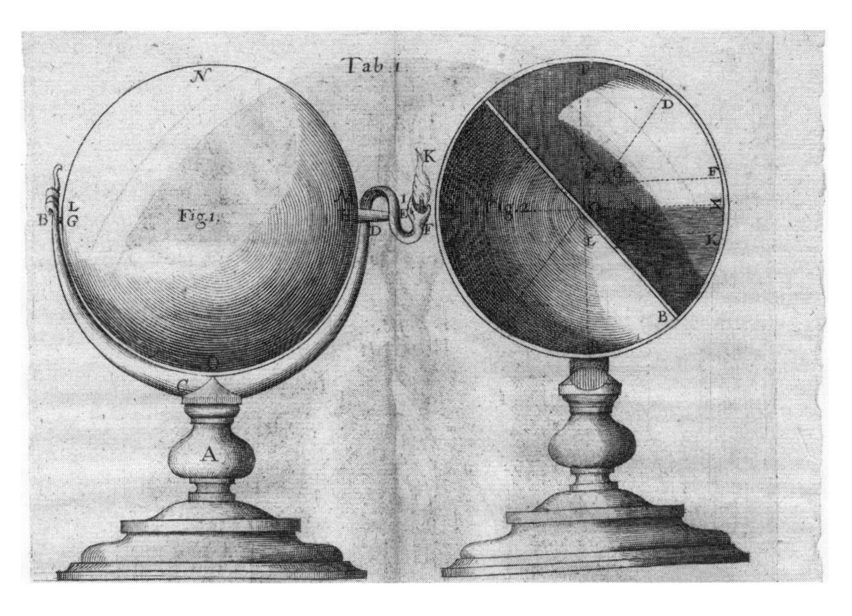

Fig. 24 Elevation and cross-section of one of Hooke's designs for lamps published in *Lampas*, 1677. (Figs. 23 and 24: Museum of the History of Science, Oxford.)

> true knowledge of the Mechanical Powers, and of a method of applying them to the Explication of Nature.[105]

Hooke himself used a telling metaphor (he was good at metaphors) to express this connection. Introducing his first published Cutlerian Lecture, he explained his intention to alternate between 'Artificial Improvements' and 'Observations of Nature': 'I design alwayes to make them follow each other by turns, and as 'twere to interweave them, being apart but like the Warp or Woof before contexture, unfit either to Cloth, or adorn the Body of Philosophy.'[106]

Astronomy and Optics

The central issue for Hooke's work in astronomical instruments was the dispute with the Polish astronomer, Johannes Hevelius, over the usefulness of telescopic sights as replacements for open-sighted alidades on measuring instruments. Hevelius had devoted both a enormous sum of money and most of a life's work to building his extensive observatory in Danzig and equipping it with a whole range of instruments. He had installed both large refracting telescopes and instruments for the traditional astronomical business of measuring stellar positions, the latter being essentially mechanical developments of the designs of Tycho Brahe. Hevelius thought telescopes were fine for observing with, but that they could not replace the Tychonic sighting arrangement, which did not, of course, use lenses.

English astronomers, such as Wren and later Flamsteed, were using telescopes with points of reference in the focal plane of the objective—that is, telescopic sights—which allowed, by this addition of 'artificial organs', a more precise alignment with the target. Anything placed in the focal plane would be magnified by the eyepiece and appear in sharp focus with the astronomical image. This could be a point, a thread or wire, single or crossed, a rule, a net, or a micrometer with wires or straight edges moved by screws. Hooke and others hoped that Hevelius would adopt this new technique and so render his observatory work even more valuable, and through Oldenburg he sent Hevelius instructions for fitting telescopic sights in May 1668.[107] Unfortunately for this ambition, Hevelius set his face resolutely against this novelty—not rules or micrometers in simple refractors, but telescopic sights on measuring instruments with divided arcs, such as large quadrants or sextants. It was in the course of this early exchange with Hooke that Hevelius claimed accuracies of a fifth or a tenth of a minute of arc, or even of a second, a sixtieth of a minute.[108] It was a claim that Hooke never forgot.

In a Cutlerian Lecture delivered at Gresham College in 1670, Hooke sought to demonstrate the advantages of a new technique that would be applied to a new generation of instruments.[109] His zenith telescope (Fig. 25) sought to capitalize on the advance in accuracy offered by the eyepiece micrometer to capture the first empirical proof of the orbital motion of the earth, which was expected to come from the detection of stellar parallax, an annual cycle of displacement predicted in the apparent positions of the stars. If the earth was moving in an annual orbit, there should be a corresponding variation in stellar measurements. It was not that Hooke or his colleagues doubted for a moment that the earth was in orbit—they held that the stars were so far away that the measurement of their parallax had been beyond previous techniques—but there was still no clinching proof.

A zenith telescope, that is, one pointing directly upwards, would avoid the disturbances of atmospheric refraction and, since good astronomical telescopes had long focal lengths to mitigate the effects of spherical and chromatic aberration, Hooke had to allow his telescope to pass through two storeys of his rooms in Gresham College. He dispensed with a tube, aligning the eyepiece by means of plumb-lines. The measurements were to be taken with an eyepiece micrometer near the floor and, although Hevelius's objections were not to micrometers but to telescopic sights fitted to instruments with divided arcs, Hooke took his opportunity to point out that, of necessity, the future advance of astronomical

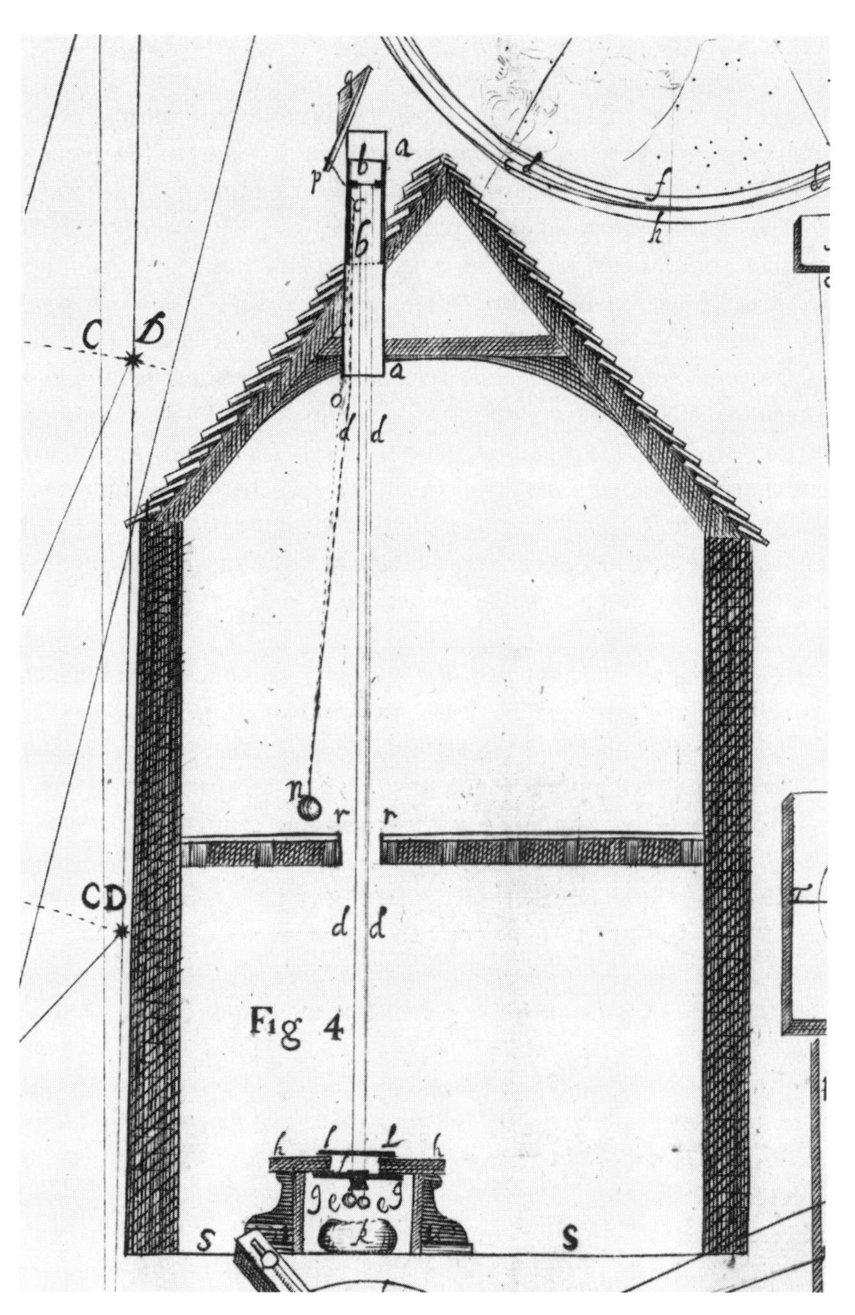

Fig. 25 Hooke's zenith telescope built into his rooms at Gresham College, with the object glass mounted in the roof and the eyepiece micrometer, close to the lower floor, aligned by two plumb-lines, from *An Attempt to Prove the Motion of the Earth* (1674). (Museum of the History of Science, Oxford.)

measurement depended on the application of sights, on account of an uncompromising physiological limit, the angular resolving power of the human eye: 'the naked eye cannot distinguish an Angle much smaller then a minute, and very few to a whole minute.'[110] This limit, he later demonstrated, extended even to the Fellows of the Royal Society.[111] Hevelius had, of course, made much more ambitious claims. Hooke makes a joke here that refers again to the idea of the decay of eyesight after Eden: 'I judged that whatever mens eyes were in the younger age of the World, our eyes in this old age of it needed spectacles'[112]—the joke being the reference to the common experience in a single human life-span, as well as in the span of human history, of coming to need spectacles.

The detailed story of the exchanges between the protagonists is complex, but for our purpose we can say that Hevelius countered in his *Machina Cœlestis* of 1673 and Hooke rejoined in his *Animadversions on the first part of the Machina Cœlestis*, based on a lecture at Gresham College in December of the same year. Two powerful emotions run through *Animadversions*, perhaps Hooke's most animated published assertion of the importance of instrumentation. One is frustration. Apparently oblivious to the personal investment Hevelius has made in a programme of observation that is about to be undermined by an unexpected new development, Hooke cannot understand why he persists in refusing to see the evident advantage of telescopic sights:

> if he had prosecuted that way of improving Astronomical instruments, which I long since communicated to him, I am of opinion he would have done himself and the learned World a much greater piece of service, by saving himself more then 1/10 of the charge and trouble, and by publishing a Catalogue ten times more accurate.[113]

In fact, Hooke does not always write as though he is completely oblivious of the plight of Hevelius, and this is where we can illustrate the second emotion—optimism for the future. Later he does acknowledge 'his great and liberal expence . . . his vast pains, care and diligence,' and that he 'hath gone as far as it was possible for humane industry to go with Instruments of that kind'.[114] His problem comes with the idea that Hevelius might stop, and stick there, and defend his position, instead of embracing whatever new development will advance his programme:

> I would not have the World to look upon these as the bound or *non ultra* of humane industry, nor be perswaded from the use and improvement of Telescopical Sights, nor from contriving other ways of dividing, fixing, managing and using Instruments for celestial Observations, then what are here prescribed by Hevelius. For I can assure them, that I have my self thought of, and in small modules try'd some scores of ways, for perfecting Instruments for taking of Angles, Distances, Altitudes, Levels, and the like, very convenient and manageable, all of which may be used at Land, and some at Sea, and could

describe some 2 or 3 hundred sorts, each of which should be every whit as accurate as the largest of Hevelius here described, and some of them 40, 50, nay 60 times more accurate . . . These I mention, that I may excite the World to enquire a little farther into the improvement of Sciences, and not to think that either they or their predecessors have attained the utmost perfections of any one part of knowledge, and to throw off that lazy and pernitious principle, of being contented to know as much as their Fathers, Grandfathers, or great Grandfathers ever did . . . Let us see what the improvement of Instruments can produce.[115]

By way of illustrating what they might produce, Hooke offers an account of a truly visionary instrument, his equatorial quadrant (Fig. 26).[116] We have to remember that at this point an equatorial motion had not been applied to an observatory instrument other than armillary spheres from Ptolemy to Tycho. Here Hooke has an equatorial mount for a full-scale astronomical quadrant. In fact the whole design is an extraordin-

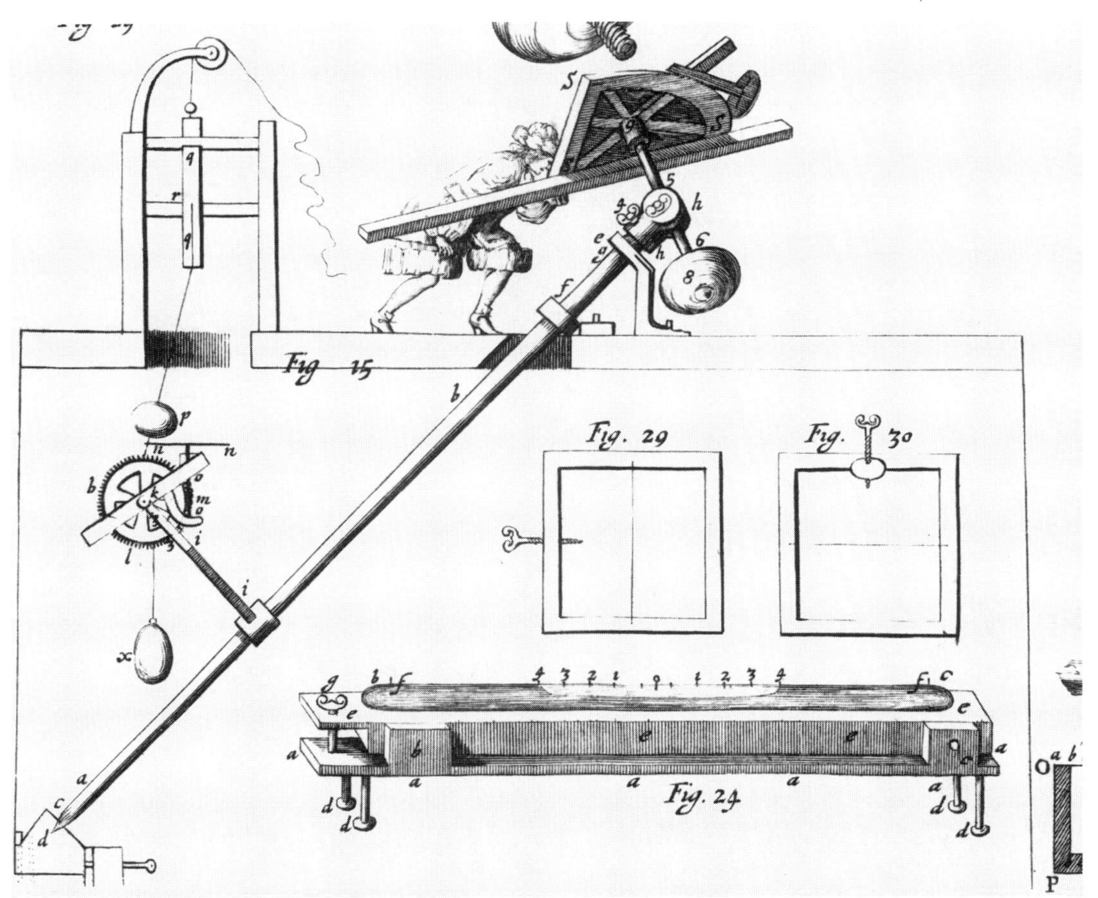

Fig. 26 Hooke's design for a large astronomical quadrant, mounted on a polar axis and driven by a conical pendulum clock. The lower figure shows his bubble level. *Animadversions on the First Part of the Machina Cœlestis of Johannes Hevelius* (1674). (Museum of the History of Science, Oxford.)

ary combination of original elements into a single audacious design. The equatorial mount itself is impressive, but Hooke is not content with that: this first such mount also has a clockwork drive. The advantage of an equatorial is that the observer can follow the motion of a target in the heavens merely by pushing the instrument around the axis, but Hooke goes further and has a machine to do the pushing. It is here that we again encounter Hooke's conical pendulum. He had earlier pointed out in connection with his conical pendulum clock that it moved 'without any noise, and in continued and even motion without any jerks'.[117] This even motion is, of course, just what is needed in a clockwork drive for an equatorial mount, where a to and fro pendulum would not be suitable, and it is characteristic of Hooke that he sees this connection, that he moves from a timekeeper of complete originality to the first motion governor for an equatorial instrument. In addition to this he has a tangent screw adjustment of the telescopic sight in its motion around the limb—this would become a common feature of future astronomical and navigational instruments. The form Hooke describes is a micrometer screw with a graduated head, engaging teeth in the quadrant limb (Fig. 27). There is a long handle from the tangent screw to the apex of the quadrant, so that the observer can move it from there, and at this apex he uses mirrors in diagonal eyepieces to superimpose the two images

Fig. 27 Illustrations of the tangent-screw adjustment and the superimposition of two images in Hooke's design for an equatorial quadrant, *Animadversions on the First Part of the Machina Cœlestis of Johannes Hevelius*, 1674. (Museum of the History of Science, Oxford.)

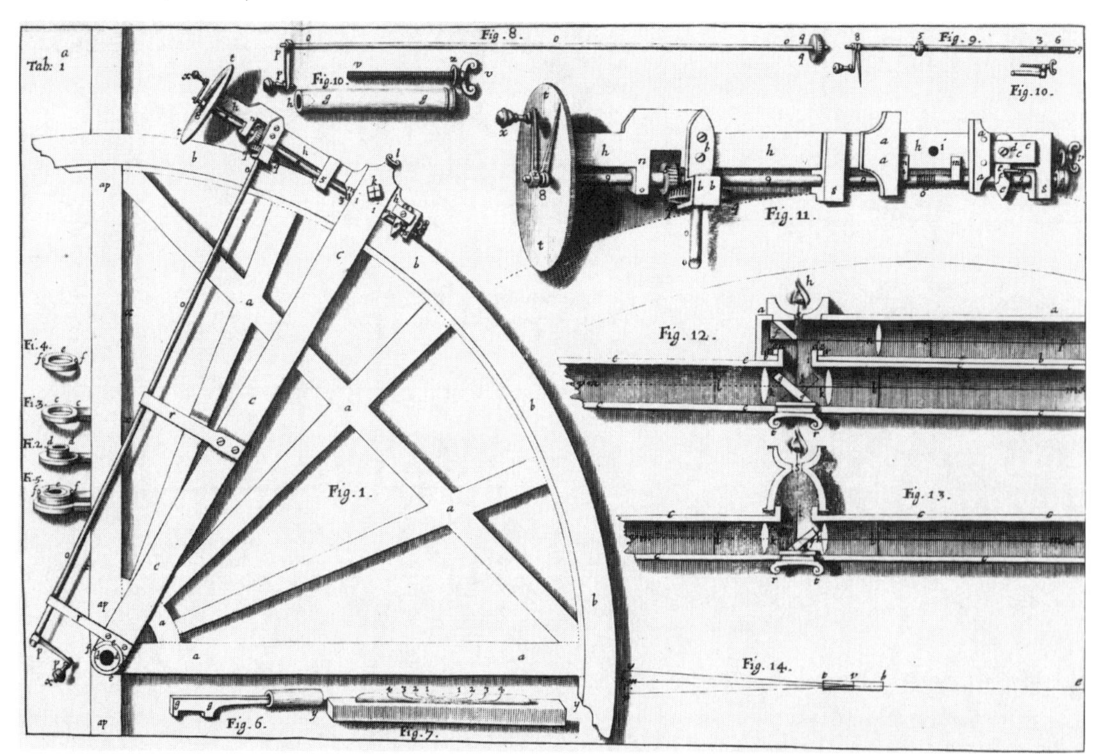

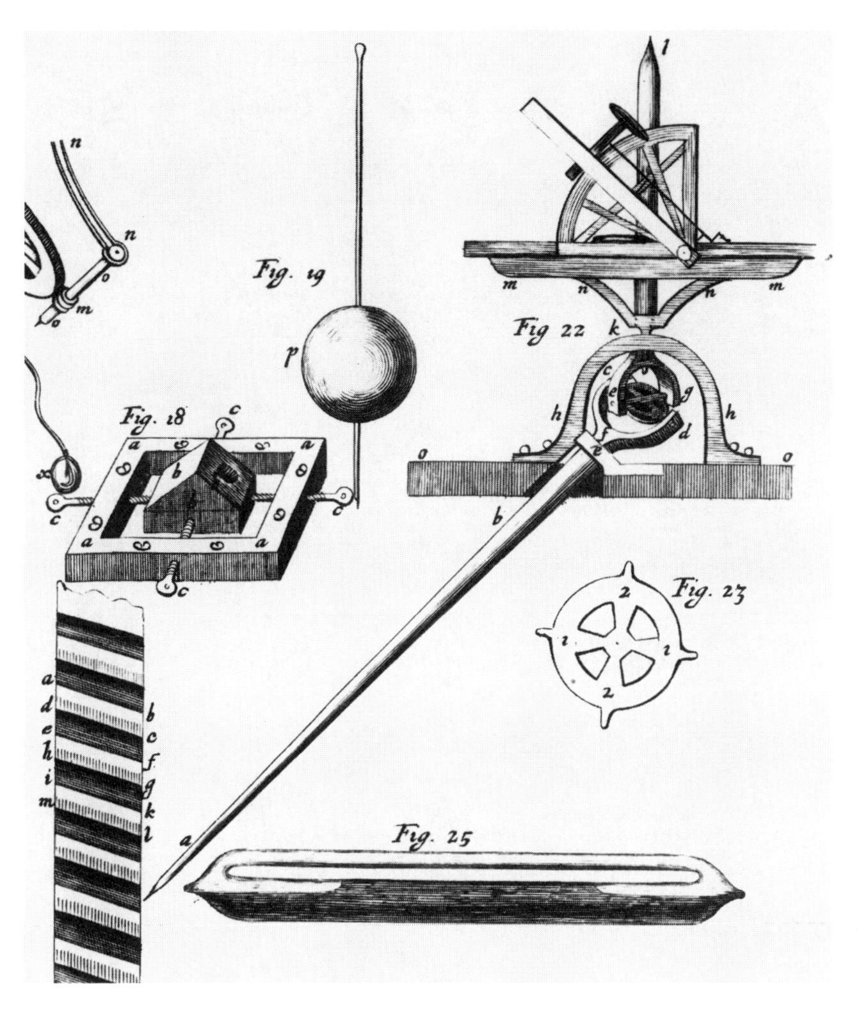

Fig. 28 Hooke's quadrant diverted into an azimuth motion using his design of universal joint, from *Animadversions on the First Part of the Machina Coelestis of Johannes Hevelius*, 1674. (Museum of the History of Science, Oxford.)

from the fixed and moving telescopic sights in order for the instrument to be used by a single observer.

One other thing to mention about this instrument is that Hooke illustrates how it may be set on a vertical axis, with the motion diverted from the equatorial drive so as to follow the heavens in azimuth instead of in right ascension (Fig. 28). To do this requires a special universal joint—the 'Hooke joint' as it is known, and much used in mechanisms to this day—which was introduced to the Royal Society in 1667 as a contrivance for laying out the hour lines on sundials set in any orientation to the motion of the sun.[118] So again we see Hooke's facility for transferring an idea between different areas of work, in this case from laying out the lines on an inclined sundial to the drive for an astronomical instrument. For the vertical adjustment he makes use of his bubble level.

Hooke has a range of other instruments[119] and micrometers[120] in astronomy and celestial navigation, though perhaps not quite the 200

Fig. 29 The 10-foot mural quadrant designed by Hooke, made by Tompion, and installed at the Greenwich Observatory in 1676. Engraving by Francis Place. (Pepys Library, Magdalene College, Cambridge.)

or 300 he boasted of above. Not all were mere projects. Hooke designed a mural quadrant that was built as one of the first instruments of the Royal Observatory at Greenwich (Fig. 29).[121] He described his methods of dividing the limb of such an instrument in *Animadversions*,[122] where he also introduced the idea of a movable subsidiary arc on a glass plate covering a few degrees only, to avoid having to extend the finest division to the whole arc,[123] a modified version of which was used in the Greenwich quadrant. Hooke's account of his equatorial quadrant certainly contributed to the design of an equatorial sextant that was installed at Greenwich in 1676.[124] One similarity is the adjustable sight moved by a

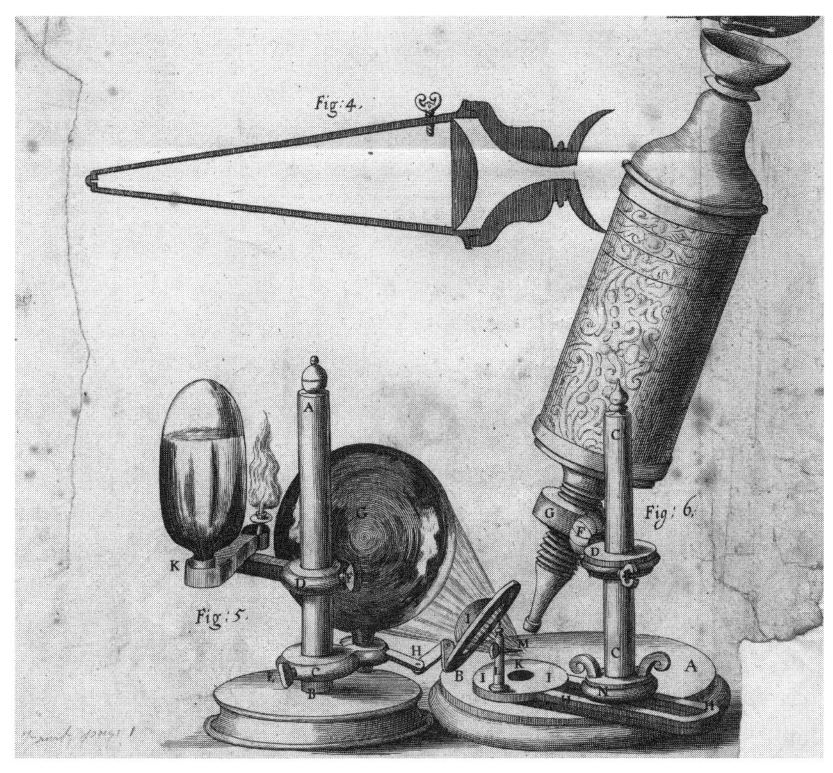

Fig. 30 The compound microscope illustrated in *Micrographia*, with an oil lamp and condensing lens for illumination. (Museum of the History of Science, Oxford.)

micrometer screw engaging teeth on the limb; this part of Flamsteed's instrument had been executed by Thomas Tompion, who had been recommended in Hooke's *Animadversions* for precisely such work.[125] Flamsteed would never have acknowledged the connection, but the antipathy there was mutual: Hooke records on 28 May 1674, 'I shewd Flamstead my quadrant. He is a conceited cocks comb.'[126]

We return in conclusion to optical instruments, to microscopes and telescopes, where, as in mechanical horology, we can benefit from the publication of a careful and thorough treatment, in this case an excellent article by Allen Simpson.[127] The microscope Hooke illustrated in *Micrographia* is a commercial product (Fig. 30), one that a customer could purchase in Richard Reeve's shop in Longacre, though there may have been a wait, since he was in demand. Hooke did some work on improving the microscope and he was well served by Reeve and later by Christopher Cock. He designed a new microscope that was made by Cock; the claim that it achieved impressively high magnification and resolution probably had substance, for it was bought from Cock by the Royal Society for its use.[128] Hooke was also keen to use simple microscopes, that is, instruments with a single lens, and to pursue the techniques associated with Leeuwenhoek, which in any case he considered he had mentioned already in the preface to *Micrographia*.[129] Also mentioned there were reflecting microscopes, and at least one was later made.[130]

Hooke claimed some originality for his methods of illuminating his subjects, either by daylight or lamplight concentrated by a condensing lens in the form of a plano-convex glass lens or a glass globular flask filled with water or brine. It is typical of Hooke's care over detail that he points out that, when using sunlight, it should be filtered through a sheet of oily paper or of glass 'rubb'd on a flat Tool with very fine sand'. If sunlight is not diffused in this way, 'the reflexions from some few parts are so vivid, that they drown the appearance of all the other, and are themselves also, by reason of the inequality of light, indistinct, and appear only radiant spots.'[131] Increased illumination offers significant advantages but Hooke looks forward to the elimination of aberration through the grinding of aspherical lenses, because although his methods do permit the application of higher powers, 'after a certain degree of magnifying, they leave us again in the lurch'.

In the case of the telescope Hooke was caught in the contemporary dilemma that improvements in optical performance—essentially improvements in aperture and light grasp, so that faint images could bear higher magnifications—could be achieved only by increasing the focal length and thus the overall length of ever more unmanageable tubes. Only with slight curvature could aberration be contained. Even then, using apertures optimally was important for squeezing the best performance from an instrument and in 1681 Hooke invented a 'new-contrived aperture for long telescopes', which sounds very like a form of iris diaphragm, 'which would open and close just like the pupil of a man's eye, leaving a round hole in the middle of the glass of any size desired'.[132] We have Hooke's own sketch of a 36-foot telescope by Reeve in a tube of 40 feet mounted in the courtyard of Gresham College, the tube being braced to prevent it bending under its own weight, and suspended from a mast by a system of pulleys and raised or lowered by a winch at the base (Fig. 31). The suspension for the pulley could move around the mast, according to the direction required of the telescope. A rest for the eye-end resembles an artist's easel, while at the foot of the page is a series of micrometers designed by Wren. Hooke writes in the accompanying account as though it is a shared resource: there had been contributions to its design by different people in his circle and it was used by other astronomers as well as himself.[133]

This was a fine instrument and Hooke often refers to an even better one of 60 feet by Reeve as a kind of standard of contemporary telescopic achievement in England.[134] It was clear, however, that the problem of increasing length had to be addressed if telescopes were to continue to

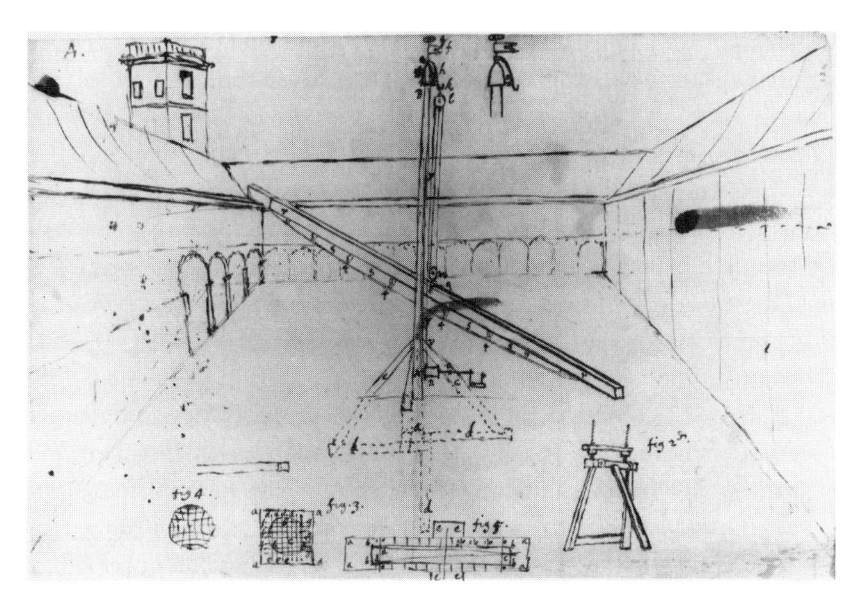

Fig. 31 Hooke's sketch of a 36-foot refractor by Reeve, mounted in the courtyard of Gresham College. (Royal Socety Classified Papers XX, no. 61.)

improve. Hooke had a number of approaches. One was to dispense with the tube, to produce the 'aerial telescope' where the observer controlled the objective mount from his station at the eyepiece, usually by means of a line.[135] The technique is generally associated with Huygens, who described it in 1684. In fact, Hooke described it to the Royal Society in 1678 and his use of it seems to go back to 1669—or 1668, if that special case of the zenith telescope at Gresham is included. Modern historians of Hooke's work espouse disinterest and objectivity, and subscribe to the belief that issues of priority are of very little concern, but it is hard at a human level not to sympathize with him about so many of these apparent anticipations.

A more complex and qualified case of anticipation concerns the advent of reflecting telescopes, and Hooke was certainly concerned with the use of mirrors in telescopes from an early date. But because Newton's reflector of 1671 was often seen as a device to shorten telescopes, and because plane mirrors in combination with an objective lens could be used to the same effect, it is not always clear whether curved mirrors were included in the early designs under discussion, or even whether anyone thought this an important distinction to make.[136] With hindsight, it seems critical to the concept of a reflecting telescope, but that may be distorting our understanding of attitudes at the time. In any case, Newton's telescope was, of course, a combination of lenses and mirrors. Hooke was showing the Royal Society his telescopes shortened by plane mirrors as early as February 1667.[137] He had experimented with the concave mirrors Reeve had made for James Gregory in 1663, in an attempt to make Gregory's reflector, so he was very familiar with the potential use of curved mirrors in the quest for shorter telescopes.[138]

As Simpson has shown, Hooke was working on Newtonian types of telescope from the beginning of 1672, and was planning to produce an instrument of 4-foot focal length, significantly larger than Newton's.[139] By the time Hooke's diary begins in earnest in August 1672, Hooke is working closely with Cock, who had previously worked for Reeve. The first month contains:

> (1) . . . Paid Coffin £2 3s. 4d for shutts, etc. . . . (3) . . . Cox turning object glasse with flints. . . . (5) . . . Coxes . . . Polisht an object speculum of 7 inches. (6) Tryd polishing all the morn. . . . Coxes founder at the ball in Lothbury. . . . (7) . . . Coxes, a little concave. . . . (8) . . . Polisht a concave well. . . . (9) . . . Club of the Society. . . . Gave Lord Brounker Reflex Speculum. (10) . . . Coxes . . . (11) Fitted my Newton. . . . Cox calld here and promisd the Specular object glasse Wensday morning. . . . (12) . . . Coffin begun the Room in the Cloysters by his man. He and his man finisht it by next noon. . . . (15) . . . Cox. . . . (16) Set up the work room in the cloyster with Coxes speculum all the morn. . . . Committee of the Society . . . (17) . . . polisht speculum. . . . Coffin made specular frame. (18) Tryd speculum hopefully. . . . (19) . . . Coxes bevelled ground speculum sent to Guy and had an end made on issue easily but the arm was afterwards stiff. . . . (21) . . . Coxes. . . . (23) . . . Pollish 9 foot speculum well. Saw Moon at night through it very big and distinct. Committee here. . . . signed Cox a bill of £5 for glasse. . . . (24) . . . Sir S. Morland liked proposition of Speculum which I told him with Grove. . . . (25) . . . With Cox after dinner about Sir S. Morelands speculum. . . . (26) With Cox at founders. . . . (29) Polisht glass. . . . (30) . . . Lent Cox the great flat speculum. Committee here . . . (31) I observd with Speculum telescope. . . . Coxes, had his plaister mould 1sh.[140]

The Royal Society had begun its summer recess on 10 July, but what Hooke calls a 'club' or 'committee' had decided to meet weekly for discussions and experiments, including 'Such, as might improve Mr. Newton's reflecting telescope: and particularly to see finished a four-foot telescope of that kind, already recommended to Mr. Cock.'[141] Hooke's record for August is both revealing and obscure. The pace is fast: a great deal seems to happen in a single month. Quite how many specula are concerned is unclear, but that Hooke was working on a mirror of 9 foot focal length shows that his private ambitions were outstripping those of the Society, which were already ambitious for the time. It also indicates that Hooke had a personal commitment to the programme: it is unlikely that he saw this work as entirely in the service of an idea that belonged to Newton.

It is instructive also that we see Hooke working on instruments with his own hands and, where appropriate, making use of artisans. Here his collaboration with Cock is particularly close, and Hooke takes a direct interest in what happens at the foundry where the metal alloy blanks are cast. The eponymous Coffin was a carpenter who did other work for Hooke; here he makes a frame that allows him to test the speculum. I have included the references to Coffin's work on the room in the cloister, in case this was specifically to provide suitable accommodation for the optical work, which the provision of shutters might suggest. The contribution of 'Guy' is unclear. Perhaps he was a metalworker of some kind; the bookseller Thomas Guy, founder of Guy's Hospital, who appears elsewhere in the diary, does not seem a convincing identification in this instance. Grove was a plasterer; whether he had any connection with the plaster mould collected from Cock a week after the meeting with Moreland is not clear.

Simpson has dealt with this episode in detail and traced Hooke's continuing work on reflectors, including some using silvered glass mirrors.[142] Hooke also maintained his interest in shortening telescopes by using plane mirrors, telescopes whose other optical components could be either lenses or mirrors (Fig. 32). By adapting the mirror surfaces, these instruments could be designed for viewing the sun, so Hooke published his methods of shortening telescopes in this way in his Cutlerian Lecture, *A Description of Helioscopes*.[143] He also worked with Cock on a shortened telescope for lunar work, his 'Selenoscope', completed in 1675.[144] We have seen that in this year he even discussed polishing techniques with Newton.

Hooke returned to the subject of long telescopes and how they might be shortened in a very late paper, surviving in manuscript at the Royal Society.[145] It is a review of Jean de Hautefeuille's method of shortening telescopes and was read to the Society in December 1697, when Hooke was 62 years old. He refers to his 60-foot telescope by Reeve, where, with eyesight sufficiently good to distinguish a half or a third of a minute at arc,

> not much more then between Broadstreet & Bishops gate street crosse this house is discoverable. that is a length of 440 foot. or. 147. yards. Now if it be possible to make a telescope that shall be at least 10 times as good as this which I believe not only not impossible but very practicable if Money were not wanting then a length of 14 or 15 yards might be Distinguished in the Surface of the moon and that would be lesse

than an animall I have seen upon the earth Namely the whale which was . . . seen at Greenwich a Little before the Death of Oliver the Protector.

This unexpected link to a distant but memorable event in the 1650s (Cromwell died in 1658) takes us right back to the preface to *Micrographia*, where in his relatively youthful enthusiasm for the potential of optical instruments Hooke had written:

> 'Tis not unlikely, but that there may be yet invented several other helps for the eye, as much exceeding those already found, as those do the bare eye, such as by which we may perhaps be able to discover living Creatures in the Moon, or other Planets, the figures of the compounding Particles of matter, and the particular Schematisms and Textures of Bodies.[146]

Hooke quotes the passage word for word in his paper of 1697. He is saying, some thirty-three years later, that he was right, he has been vindicated, and that he has not changed his mind—that he still believes in the extraordinary potential of future instruments. As well as reviewing Hautefeuille, he takes in the opinions of Adrien Auzout and Nicolaas Hartsoeker on the limits of telescopes and their scepticism about Hooke's

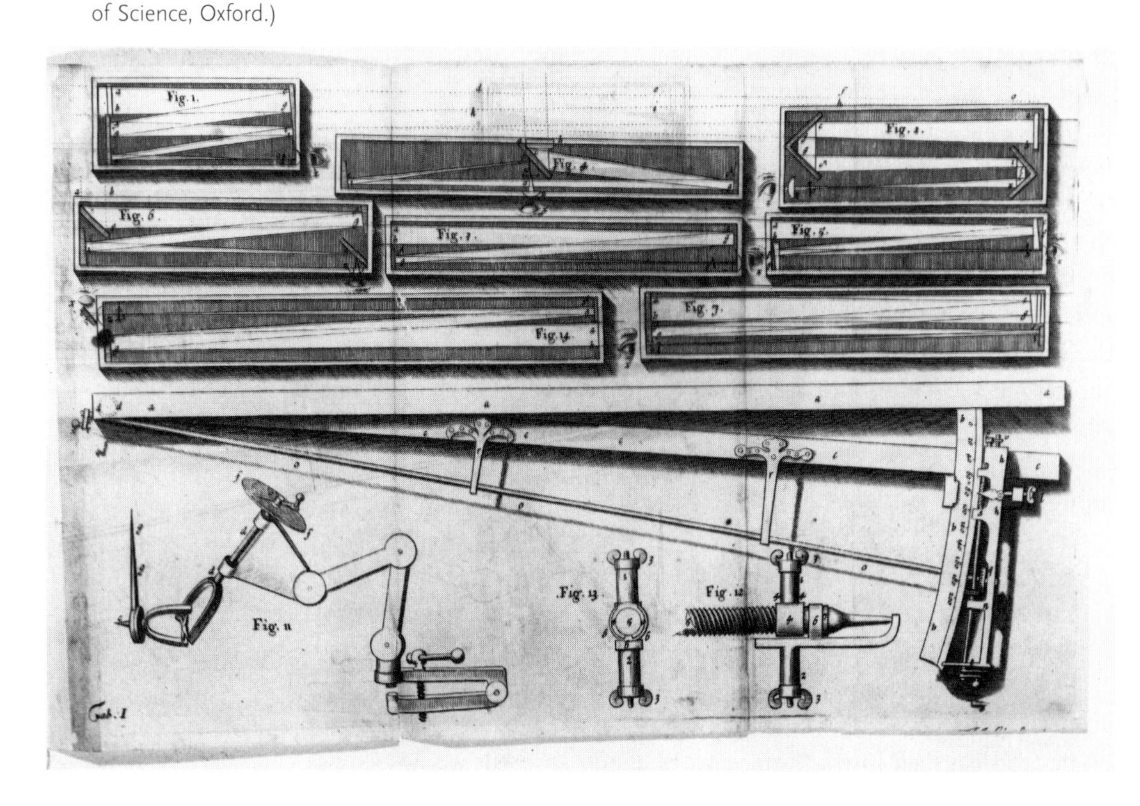

Fig. 32 At the top is a variety of optical configurations devised by Hooke, in which mirrors are used to shorten telescopes. The bottom left illustration is of the Hooke joint applied to laying out the hour lines on sundials. From *A Description of Helioscopes* (1676). (Museum of the History of Science, Oxford.)

claim that animals might one day be seen in the moon. His dry tone shows Hooke's wit intact at the age of 62.

Hartsoeker lays claim to a successful object glass of 600 foot focal length, which will allow an aperture of one foot and an eyepiece of 1 foot focal length. This will reveal a spot on the moon of 3,500 feet diameter subtending an angle of six minutes. This, says Hartsoeker, is the limit of achievable resolution, a limit which Hooke characteristically challenges immediately:

> we are not to hope to be able to goe any farther and what pray is his Reason, why tis because of the incomodiousnesse and Difficultie that Long telescopes are subject to by Reason of the Rapid motion of the Starrs.

Of course, Hooke still rejects arguments from present impossibility, and in the discussion of this he clearly imagines that some wholly new approach to telescopic optics will yield instruments altogether different in principle, in contrast to his antagonists, who simply imagine the extension of the existing techniques into ever longer telescopes.[147]

Here Hooke cannot resist a jibe at his old opponent, Hevelius, though he had been dead for ten years. With the telescope in question a sharp-sighted man, able to distinguish half or a third of a minute—'which may be possible'—will be able to distinguish a spot of 194 feet on the moon, and with a telescope that can magnify just over 32 times as much this sharp-sighted man will see an object 6 feet in diameter—'and such a telescope as this I doe not at all look upon as impossible'.

> But if we has such a sharp sight as Mr. Hevelius who affirmes to be able without glasses to distinguish seconds then such a one would need a telescope noe better then this of 600 foot nor indeed need it be altogether soe good . . . with one a tenth part as good such an Hevelian sighted man would be able to see a whale of 60 foot Long & that I take to be an animall.
>
> And thus setting our shoulders to it . . . we have removed the Herculean Columns . . . and have plucked up and carryed of the Posts Gates and barrs of those Gates of Gaza where these 3 gentlemen would have us believe we were imprisoned. . . .
>
> I am not therefore for limiting or stinting the boundaries of Art and the industry of Man. many other Ways and means may be thought of and Perfected then what the Present Generation are posessed of and there may be yet Space and Room for Succeeding generations to Imploy their thoughts and Industry in further invention & Discovery.[148]

Among several general conclusions and impressions we should take away from a review of Hooke's instruments, one is his optimism and confidence for the future. Another is the coherence and integrity of his programme across technological devices, mechanical and mathematical instruments, and his practice of the experimental mechanical philosophy. We have seen a certain justice in his complaints of a lack of appreciation and acknowledgement, while at the same time we must assent to the customary caveats about his being less skilled at carrying through and systematizing his work. We have also seen the considerable influence of his celebrated fecundity of ideas and diversity of fields of practice. Often he carries devices across the boundaries that limited other practitioners —taking magnetics into horology, conical pendulums into instrument mounting, the universal joint from sundials to instrument drives. Depth sounders become probes for investigating the deep sea, hygroscopes stimulate speculation about muscular action, hydrometers about planetary dynamics, and so on.

We should end with *Micrographia*—the outcome of Hooke's most famous engagement with instruments—but so as to show that it too reveals a Hooke of a different stamp from the usual caricature. He concludes the preface with a delightful conceit and a generous gesture. With reference to the parable told in the New Testament by Jesus, a widow contributes a mite—a small coin—to the treasury of the Temple in Jerusalem, but it is of greater worth than the much larger sums offered with ostentation by more wealthy donors, because it is given out of poverty and with a sincere heart. Hooke offers his mite—in his case a microscopic insect—to the treasury of philosophy; it is a touching metaphor, which he follows by using his tiny insects to construct a charming compliment to the more weighty work of other natural philosophers:

> I have at length cast in my Mite, into the vast Treasury of A Philosophical History. And it is my hope, as well as belief, that these my Labours will be no more comparable to the Productions of many other Natural Philosophers, who are now every where busie about greater things; then my little Objects are to be compar'd to the greater and more beautiful Works of Nature, A Flea, a Mite, a Gnat, to a Horse, an Elephant, or a Lyon.[149]

Notes

1. Aubrey 1898, i, 415.
2. Royal Society Classified Papers XX, no. 78.

3. Bennett 1997, 30.
4. Hooke 1665, preface, sig. a1r.
5. Ibid.
6. Ibid., sig. a2r.
7. Ibid., sigs a2r–a2v.
8. Hooke 1679, 65.
9. Birch 1756–7, iii, 364–5.
10. Hooke 1665, preface, sigs b2v–c1r.
11. Ibid., sig. c2r.
12. Ibid., sig. c2v.
13. Ibid.
14. Ibid., sig. a2r.
15. Ibid., sig. b.
16. Ibid., sig. a2v.
17. Wright 1989.
18. Ibid., 77.
19. Ibid., 98.
20. Hooke 1935, 148.
21. Ibid., 150.
22. Ibid., 148.
23. Birch 1756–7, ii, 112.
24. Hooke 1935, 157.
25. Ibid.
26. Ibid.
27. Hooke 1679, 207–8.
28 Birch 1756–7, ii, 83, 350, 355–6, 359, 360, 372, 374.
29. Ibid., 372.
30. Ibid., 388.
31. Ibid., 361, 388.
32. Ibid., 398.
33. Hooke 1705, 545–6.
34. Howse 1970–1; Howse 1975, 125–6.
35. Gunther 1930a, 314; Birch 1756–7, ii, 203–4.
36. Hooke 1705, p. ix; Birch 1756–7, i, 417.
37. Birch 1756–7, ii, 97, 105, 137, 139; Hooke 1679, 69, 105–6; Hooke, 1705, p. viii.
38. Birch, 1756–7, ii, 150–1, 153, 168.
39. Royal Society Classified Papers XX, no. 53; Pugliese 1989.
40. Birch 1756–7, ii, 90–2.
41. Hooke 1679, 105–6.
42. Birch 1756–7, i, 370.
43. Birch 1756–7, iii, 191; Hooke 1935, 149.
44. Birch 1756–7, ii, 113.
45. Hooke 1705, 503; Birch 1756–7, ii, 447.
46. Birch 1756–7, ii, 128; Hooke 1705, 557–8.
47. Hooke 1705, 557; Sprat 1667, 246.
48. Birch 1756–7, i, 287, 297.
49. Ibid., 197, 307, 316, Royal Society Classified Papers XX, no. 23.
50. Birch 1756–7, i, 330.
51. Ibid., iii, 397–8.
52. Ibid., i, 287, 297, 307–8; ii, 133, 134, 136–7, 139, 385.
53. McConnell 1982, 6–11.
54. Birch 1756–7, iii, 394, 395–6, 399.
55. McConnell, 9–10.

56. Ibid., 10.
57. Ibid.
58. Hooke 1705, 561–2; Birch 1756–7, iv, 230–1.
59. Birch 1756–7, i, 180–2; iii, 394, 395–6; Hooke 1726, 225ff.
60. Middleton (1964) regards Pascal as the inventor of the siphon barometer, but since he did not publish the idea, the work of Hooke and Boyle was not derivative from him. For Hooke's portable siphon barometer, see Birch 1756–7, i, 465–6.
61. Birch 1756–7, ii, 235, 239; *Philosophical Trans.*, xxii (1700–1), 791–4; Hooke 1705, 554–7.
62. Turner 1983, 232.
63. Birch 1756–7, i, 365, 367, 371; iii, 409–10; Hooke 1665, Preface, sigs c1v–c2r; Sprat 1667, 173; Hooke 1726, 302; Royal Society Classified Papers XX, no. 32; *Philosophical Trans.*, i (1665–6), 218–19; Boyle 2001, ii, 343; iii, 119–20; Middleton 1964, 94–9.
64. Birch 1756–7, ii, 298; *Philosophical Trans.*, xvi (1686–92), 241–4; Middleton 1964, 88–93.
65. Hooke 1665, 38–9; Hooke 1705, 555–7; Birch 1756–7, ii, 5; iv, 72.
66. Birch 1756–7, iii, 476–7; Hooke 1726, 43.
67. Sprat 1667, 173; Birch 1756–7, iv, 223, 225–6.
68. Birch 1756–7, i, 467; ii, 1; iii, 73, 78, 222, 432, 445, 450, 453, 476–7, 486, 487–8; iv, 277, 283; Hooke 1726, 41.
69. Birch 1756–7, i, 271, 311, 320; iii, 478, 479–80, 481.
70. Hooke 1665, 150.
71. Birch 1756–7, i, 271–2.
72. Hooke 1665, 151–2.
73. Ibid., 152.
74. Hooke 1705, iii–iv; Boyle 1999, i, 143–301, especially 159–63; Birch 1756–7, i, 214, 380; Shapin and Schaffer 1985, 26–30, 231–5.
75. Birch 1756–7, i, 125, 219, 234–6, 250, 255, 322; iii, 446.
76. Ibid., iii, 344, 348, 364.
77. Ibid., i, 443, 453, 459, 463–4, 473, 480, 485, 489; iii, 402; Boyle 2001, ii, 343–4, 399; Hooke 1665, Preface, sigs e2r–f1r.
78. Hooke 1705, 127.
79. Birch 1756–7, ii, 417.
80. Royal Society Classified Papers XX, no. 10.
81. Birch 1756–7, ii, 49; iii, 131; iv, 494; Hooke 1705, 486.
82. Birch 1756–7, ii, 72, 75; iii, 114; iv, 66.
83. Hooke 1705, 484.
84. Birch 1756–7, iii, 114; Hooke 1935, 74.
85. Birch 1756–7, iii, 115.
86. Ibid., i, 443, 449, 455–6, 460, 461; Hooke 1705, 550; Gunther 1930a, 200–1.
87. Birch 1756–7, i, 179, 192, 193–4, 195–7; Hooke 1705, 16; Royal Society Classified Papers XX, no. 12.
88. Birch 1756–7, i, 197.
89. Bennett 1982, 71–2.
90. Hooke 1679, 152, 337–8.
91. Birch 1756–7, i, 329, 333, 334, 337, 348; ii, 436, 440, 442, iv, 103, 105, 113, 115, 118, 120, 132, 408, 420, 498; Hooke 1705, 519, 530, 533–5; Gunther 1930b, 753–6, 758.
92. Hooke 1726, 296; Birch 1756–7, ii, 403.
93. Birch 1756–7, ii, 483, 484, 492.
94. Ibid., ii, 128, 169, 306; iv, 34, 36, 347, 483–4.
95. Birch 1756–7, i, 292, 295, 297, 302–3, 335, 342; Royal Society Classified Papers XX, no. 22.
96. Birch 1756–7, iii, 73, 77, 85–7; Hooke 1679, 81; Hooke 1705, p. xix.
97. Birch 1756–7, ii, 208, 235, 240, 258, 500.

98. Ibid., 436.
99. Ibid., i, 332, 334.
100. Hooke 1726, 107; Hooke 1679, 152; *Philosophical Collections*, no. iii (1681), 61–4.
101. Birch 1756–7, iv, 113, 117.
102. Ibid., i, 417, 477, 483; ii, 52, 256, 377, 379, 385, 416, 463.
103. Ibid., ii, 146, 155–6; iii, 229–30, 231, 247; Hooke 1679, 156ff.; Royal Society Classified Papers XX, no. 47.
104. See also Birch 1756–7, ii, 259.
105. Hooke 1705, ii–iii.
106. Hooke 1679, preface to *An Attempt*.
107. Oldenburg 1965–86, iv, 393–8.
108. Ibid., v, 241, 244; for the details, see Bennett 1989.
109. Hooke 1679, 17–22.
110. Hooke 1679, 9.
111. Birch 1756–7, iii, 120, 121.
112. Hooke 1679, 9.
113. Ibid., 38.
114. Ibid., 79–80.
115. Ibid., 80–1.
116. Ibid., 1679, 81–114.
117. Birch 1756–7, ii, 97.
118. Ibid., ii, 156, 158–60; Hooke 1679, 133–41; he also applied it to a lathe, Hooke 1679, 143.
119. Birch 1756–7, i, 463, 466, 468, 495, 496, 503, 507; ii, 18, 58, 69, 111, 114, 315, 362, 433, 435, 491, 492; iii, 121, 127, 133, 218, 331; iv, 102, 459, 461, 463, 465, 467, 468; Hooke 1705, 361–2, 508–9; Hooke 1726, 206; Boyle 2001, ii, 493–4.
120. Birch 1756–7, ii, 187–8, 188–9, 199, 204, 210–11, 237, 400, 409; Hooke 1679, 19–22; Hooke 1705, 496–7, 498.
121. Howse 1975, 17–19.
122. Hooke 1769, 50–2.
123. Ibid., 49–50.
124. Howse 1975, 75–9.
125. Hooke 1679, 90.
126. Hooke 1935, 105.
127. Simpson 1989.
128. Birch 1756–7, iii, 418; Hooke 1935, 365.
129. Birch 1756–7, iii, 110, 358–9, 364, 393, 437; iv, 104; note also ibid., ii, 358; Hooke 1679, 312–15.
130. Simpson 1989, 45.
131. Hooke 1665, Preface, sig. c1r.
132. Birch 1756–7, iv, 96.
133. Royal Society Classified Papers XX, no. 61.
134. Oldenburg 1965–86, ii, 387; iii, 348–9; Gunther 1930a, 279–81; the Royal Society commissioned a 120-ft lens from Reeve in March 1667, Birch 1756–7, ii, 158.
135. Birch 1756–7, iii, 388–9, 390, 392; iv, 308; Royal Society Classified Papers XX, nos 62, 91; Simpson 1989, 57.
136. Simpson 1989, 48–9.
137. Birch 1756–7, ii, 152, 158, 174.
138. Simpson 1989, 44–5.
139. Ibid., 49.
140. Hooke 1935, 4–6.
141. Birch 1756–7, iii, 57.
142. Simpson 1989, 49–54; note also Birch, iv, 532.

143. Hooke 1679, 121–9; Birch 1756–7, ii, 180; iii, 179, 220, 224.
144. Simpson 1989, 56; Hooke 1679, 152.
145. Royal Society Classified Papers XX, no. 91b.
146. Hooke 1665, Preface, sig. b2v.
147. One idea he had was the potentially promising technique of mounting separate lenses in cells and filling the space between with liquids with different refractive properties. Simpson 1989, 45; *Philosophical Trans.*, i (1665–6), 202–3; Gunther 1930a, 196–7, 239–40.
148. Royal Society Classified Papers XX, no. 91b. Paragraph breaks have been added for the benefit of the reader.
149. Hooke 1665, preface, sig. g2v.

3

Hooke the Natural Philosopher

MICHAEL HUNTER

ON 26 JUNE 1689, Robert Hooke, then aged 53, rose to address a select audience of virtuosi at the Royal Society (Fig. 33). It was early on a Wednesday afternoon, and as usual Hooke was giving a Cutlerian Lecture prior to that day's meeting of the Royal Society. The phenomenon that he was demonstrating was the strange effect that resulted from the mixture of two liquids—water and oil of vitriol, in other words, concentrated sulphuric acid—'which immediatly produced a considerable Degree of Heat'. He went on to explain how this was evidently because 'the two liquors when mixed Did coalesce and incorporate into a Lesser space then they had both possessed when separate and Distinct', and to indicate the significance and corollaries of this phenomenon. But this lecture, like many that Hooke gave to the Royal Society in the course of his career, and not least in his later years, is interesting for two further components. One is a lengthy gripe about how his ideas had been alternately undervalued and plagiarized by others. As he wrote:

> I have had the misfortune either not to be understood by some ‹who have asserted I have done nothing› or to be misunderstood or misconstrued (for what ends I ‹now› inquire not) by others ‹who have› secreatly suggested ‹that› their expectations ‹how unreasonable soever› were not answered, yet an Impartiall man may find that these have proceeded from some other prejudices & will have ‹Reason to think that those persons would have› done more service to philosophy if they ‹had› forborn finding faults with others performances and indeavoured

Fig. 33 The frontispiece of Thomas Sprat's *History of the Royal Society* (1667), a striking image designed by John Evelyn, which depicts a bust of Charles II being crowned by an angel denoting fame, flanked by the Society's first President, Lord Brouncker, and Francis Bacon, the Society's inspiration, who had died in 1626. Under the portico, which bears the Society's coat of arms, are various books, instruments, etc., associated with the Society's enterprise. From the presentation copy to the Royal Society. (Royal Society)

> to ‹have› done somewhat ‹more› themselves ῥᾶον μωμᾶοδας ἢ μιμεῖοδας tis easyer to blame others then to doe things that are praiseworthy. And though many of those things I here first Discovered could not find acceptance. Yet I finde there are not wanting some who pride themselves in the arrogating of them ‹for› their own.[1]

In this case, he went on to accuse the physiologist, John Mayow, of plagiarizing his initial discovery of the phenomenon to which the presentation was devoted.[2] But comparable complaints recur in many such lectures, not least concerning Hooke's arch-enemy, Newton, and 'those proprietys of Gravity which I myself first Discovered and shewed to this Society many years since, which of late Mr Newton has done me the favour to print and Publish as his own Inventions'.[3] Equally striking is the opening of the lecture, which gives a memorable general account of Hooke's and the Royal Society's intellectual agenda. This is worth quoting in full, since this lecture (like many from Hooke's later years) has never been published:

> The Design of this Society being for the Improvement of Naturall Knowledge that is for the Ascertaining and increasing the knowledge of the Powers and operations of Nature ‹or Naturall bodys›, I have made it my indeavour in all my Inquirys and Disquisitions
>
> Either ‹first› to define and Reduce to a ‹geometricall› certainty the Powers and effects ‹of Naturall bodys› already in part known, by Stating and Limiting them And their proper extents according to Number Weight and Measure
>
> or Els Secondly to Discover some new proprietys Qualifications or powers of Bodys not before taken notice of, by meanes whereof there might be Administred to an inquiring Naturalist A new medium or meanes to Discover the true essence and Nature of that body
>
> Or thirdly to invent and Exhibit some new artificiall ways and Instruments to inable such as should think fitt to use them, to make more curious and deeper searches into the nature of Bodys and their Operations
>
> Or 4^{ly} to invent and search out, by a proper ‹Synthetick› method of Reasoning from effects to causes such theorys as to me seemed capable to give an Instructive Direction for further Examination by experiments; such consequences as by Analyticall Resolution seemd to be the necessary Results of such a theory and conclusion
>
> Or in the 5th place to proceed with the ‹further› examination of such Experiments in order to the confirmation of the Doctrine propounded if they answered In all particulars to the effects that were expected or to the Amending Limiting and further Restraining therof if somewhat new and not expected occurred thereupon
>
> Or sixthly to Collect such Observations as are Recorded in Naturall Historians and philosophick writers as might give farther confirmation or information concerning the present Inquiry
>
> Or seaventhly to produce such geometricall Demonstrations of asserted proprietys as put the Doctrine beyond further Dispute.

In a sense, this sets the agenda of this section of the book, in which we come to the component of Hooke's life that has so far been largely missing, namely his achievement as a natural philosopher—a theoretician of science as much as a practitioner of it. Of course, this overlaps with what Jim Bennett has written above about Hooke as a purveyor and user of scientific instruments: as Bennett makes clear, and as he illustrated in a seminal article published many years ago, the use of instruments was central to Hooke's science, and there are times when it is as if it was through his instruments that Hooke formulated his understanding of the workings of nature.[4] Indeed, the relationship between 'art' and 'nature' is something of a leitmotif of his thought, in terms of both the understanding and exploitation of the natural world, and I shall return to the significance of this for Hooke in due course. But it is important to do justice to him as a natural philosopher in the fullest sense of that phrase, one whose fertile intellect produced endless speculations about the nature of the universe, and who also theorized in an interesting way about the method of science. Indeed, from either point of view, Hooke stands comparison with any of his contemporaries, and it would not be going too far to claim that, in many respects, he epitomizes the potential and the achievements of the science of his day.

He is especially important because of his central role in the Royal Society in its formative years, that seminal institution founded to champion the reform of natural philosophy by collaborative empirical enquiry. Arguably, as Curator of Experiments to the Society, Hooke did as much as anyone else to define the Society's scientific programme, while it was also in this context that he formulated his general thoughts on the aims and desiderata of science. Yet at the same time, there are certain rather paradoxical elements about Hooke, many of which are also typical of the science of his day, even if he may exemplify them in an extreme form. Indeed, doing justice to this helps to put certain critical evaluations of Hooke in context, perhaps particularly that of R. S. Westfall enshrined in the *Dictionary of Scientific Biography*, which downgrades Hooke by comparison with Newton.[5]

One is that, for all the enthusiasm that Hooke shared with other scientists of his day for reforming knowledge by providing a solid basis of empirical data carefully collected and systematized, he was also endlessly fertile in coming up with explanations of how the world worked, often of quite a speculative nature, and the urge to explain sometimes outran its empirical basis. There are also issues to do with the kind of explanations to which Hooke had recourse in his attempts to understand the workings of nature—the extent to which his thought juxtaposed what might be called magical elements with his commitment to the mechanical philosophy—which I also want to address here. Lastly, it is impossible to avoid the issues of intellectual personality that arise in an acute form in a figure like Hooke, as epitomized by the complaints that I have already cited. Hence I will end with some broad reflections on such attitudes and their context, thus helping to give a proper perspective to the dispute with Newton which has had such a blighting effect on Hooke's reputation.

The Reluctant Author

At the outset, there is an important point to be made about the material on which I will be basing my account, which has implications for assessing Hooke's intellectual achievement. It may sound paradoxical to say this, since Hooke has some famous books to his credit, notably *Micrographia*, but a case can be made for seeing Hooke as a rather reluctant author, who never produced a book unless he had a good reason for doing so. There is no example of his spontaneously tossing off a book almost for the sake of it, as might seem to be the case with his mentor, Robert Boyle, who brought out over forty books in the course of his career, with no 'careerist' incentives to do so in that, as an aristocrat, he could have emulated those of his peers who devoted themselves to hunting and fishing. Neither did Hooke ever produce a systematic treatise on natural philosophy, as found in the writings of authors like Descartes or Newton; even Hooke's most systematic work, *Micrographia*, represents something of a rag-bag by comparison, and the rest of his writings are more haphazard still.

In fact, virtually all Hooke's writings relate to his contractual obligations. Here I need to build on one aspect of what Michael Cooper has said about Hooke's patronage and his sources of income, namely his employment by the Royal Society as Curator of Experiments and the Cutlerian Lectureship that he held from 1664 until his death, paid for by the endowment of Sir John Cutler. These posts required him to provide a series of presentations which stimulated him to compose and preserve many of the writings on which this section of the book will be based. (To a lesser extent, this also applied to the Professorship of Geometry at Gresham College that he held from 1665 onwards, about which I will also say a little here.)

From the period of his life prior to his entering into these obligations, virtually no writings by Hooke survive, either in manuscript or printed form: the only exception to this is his *Attempt for the Explication of the Phænomena, Observable in an Experiment Published by the Honourable Robert Boyle Esq.* of 1661 (Fig. 34), which in a sense proves the rule because of its obviously career-advancing role. Otherwise, we are much in the

An
ATTEMPT
FOR THE
EXPLICATION
Of the
PHÆNOMENA,
Obſervable in an Experiment Publiſhed
by the Honourable
ROBERT BOYLE, Eſq;
In the XXXV. Experiment of his Epiſtolical
Diſcourſe touching the *AIRE*.
In Confirmation of a former Conjecture
made by *R. Hooke*.

Nos cum non ſemper magna referre poſſimus, vera tamen ſed rara recitamus; neq;-enim minori miraculo in parvis Natura ludit quam in magnis, Cardan de Vari. *L.8. Cap. 43.*

Tum vero de Scientiarum progreſſu ſpes bene fundabitur, quum in hiſtoriam naturalem recipientur & aggregabuntur complura experimenta, quæ in ſe nullius ſunt uſus, ſed ad inventionem cauſarum & ax omatum tan um faciunt, Verulamii Nov. Org. Aph. 99.

LONDON,
Printed by *J. H.* for *Sam. Thomſon* at
the Biſhops Head in St. *Pauls*
Church-yard, 1661.

Fig. 34 The title-page to Hooke's *Attempt for the Explication Of the Phænomena* (1661). (Museum of the History of Science, Oxford.)

dark about the formative periods of Hooke's intellectual evolution, frustratingly dependent on little scraps of information in the retrospective biographies of him by Aubrey and Waller.

From 1662, however, Hooke's requirement to fulfil his obligations stimulated him not only to the public performance of experiments and lectures, but also to the preservation of the ideas he put forward in written form and in some cases to their publication. The issue of record is important, since many extant lectures are endorsed with a note of when they were delivered and who was present, evidently as evidence that Hooke had indeed fulfilled his contractual obligations to Cutler and to the Gresham Trustees (Fig. 35). (One presumes that many of the lectures

Read before the R. Society July. 10. 1689. Present Sr J Hoskins. Mr Hill. Mr Henshaw Mr Aubrey Dr Slone. Dr [illegible]
Dr Mills. Mr S George Ash. Dr Harwood - Mr Aubrey. Mr Holly. Mr Waks. Mr Hunt. Sr J H. said the Society would willingly be at
the charge of any thing I should propound. and yt if I did not doe all I was able I should be the occasion of ye Societys Ruine
I answerd I would doe as much as any to uphold it & more than I was obleigd to. &c -

I did the Last day wth Submission propound some things which I conceived were very desirable to be
effected and such as I judged did Lye within the Limitts of possibility. for that I conceive with
my Lord Verulam that all those things may well be supposed possible and performable
which may be accomplished by some one person, though not by every one, or which may
be done by the united Labours of many though not by the single Labour of any one by himself
or which may be effected in Severall ages though possibly not in one. or what may be finished
by the care and charge of the Publique though not by the Abilitys and Industry of Private
Persons. what therefore is necessary or at Least very desirable and of good use if obtaind.
when it Lyes within the Limits of possibility ought to be attempted in those ways at Least where there
Seems the most probability. though it oft times happens too that many things are Discovered by
persons and by ways where there seems the Least probability but they are accidentall
and not to be relyed upon. However all the ways of Possibility may Lye sometimes
within the Reach of some one, who by his own abilitys and Interest may be able to
excite & Sett all the other possible agents at work. where the nature of the thing will con-
forme to the Limitts of time. but where things cannot be hastned As in observations of Comets or meteors or such like which are ye
production of Nature and fall not within the Comand of Art. to prescribe or Limitt their
appearances or Actings. there desires and art must cease. But Art affords scope enough for usefull and Necessary Disquisitions. And those other phænomena of Nature which are but Rare, are generally Lesse conside-
rable if we respect the increase of Such a Scientificall naturall knowledge as tends to practise.
There are quæsita enough to employ all that are willing to search and to enquire into the Nature
of Bodys within our power, of which as yet possibly we know very little as we ought; that is Deter-
minately Experimentally & Scientifically. for instance Even the Air Though it may seem to have been
exhausted even to a vacuity, may yet be found a plenum of other quality, vertues, uses, or powers than those we have hitherto touched upon or thought of. the like may be sayd of the water. And much
more of the earth & Earthy bodys. And if we Comprise Mineralls, metalls, Stones, &c.
it seems boundlesse. What have we yet concerning the Nature & vertue of Plants but
what is meerly Conjecturall, unless it be the naming and description of their outward form & the application of some parts of them to
Some mechanicall uses, or such uses as Nature it Self has dictated for the food of Animalls,
But as to their vertues or qualitys for Medicine or for many other physicall & mechanicall uses we are
yet in the Darke and the utmost that is known concerning them seems to be very superficiall and very uncertaine, if compared with what is yet unknown which I Conceive
Lyes within the Limitts of possibility to be detected & made evident
Againe, if we inquire for the Quæsita or the Desideranda about the Parts of Animalls. And above all the rest concerning the true formes of the parts of Men
the qualification of the humors and the use of All the more solid as well as the more
fluid parts we shall find that Even this part of Science which seems to have been the
most Cultivated and the most examined of any, because of the future reward & be-
nefit that is Expected from such a knowledge; is yet imperfect and that yet there are many parts of this thick
& Spongy wood or forrest, that have never been passed or Seen, which the many Late Discove-
rys that have been made will make somewhat the more probable. but the Pre-
valency of Diseases notwithstanding all the Drs Recipes will yet more make it
Evident. The discovery of all wch I suppose to ly within the Limits of possibility

Fig. 35 The manuscript of the Cutlerian Lecture delivered by Hooke on 10 July 1689, endorsed with a note of who was present, and a record of an exchange with Hooke's friend, Sir John Hoskins. (Royal Society Classified Papers XX, no. 78, folio 172v; see Hunter 1989, 336.)

published by Waller in his edition of Hooke's *Posthumous Works* were similarly endorsed, but that he saw such information as trivial and therefore did not record it before disposing of the manuscripts in question.)

The relationship with Hooke's obligations is equally obvious in his record of publication. As we shall see, *Micrographia* was as much a Royal Society book as Hooke's own, since it was seen by the Society as exemplifying the empirical method by which it set such store, forming one of a number of more or less effective printed manifestos that the Society authorized in its first decade. This explains why the Society felt concerned about the contents of the book, and its balance between observational and theoretical material, which we will come to later. It is important to

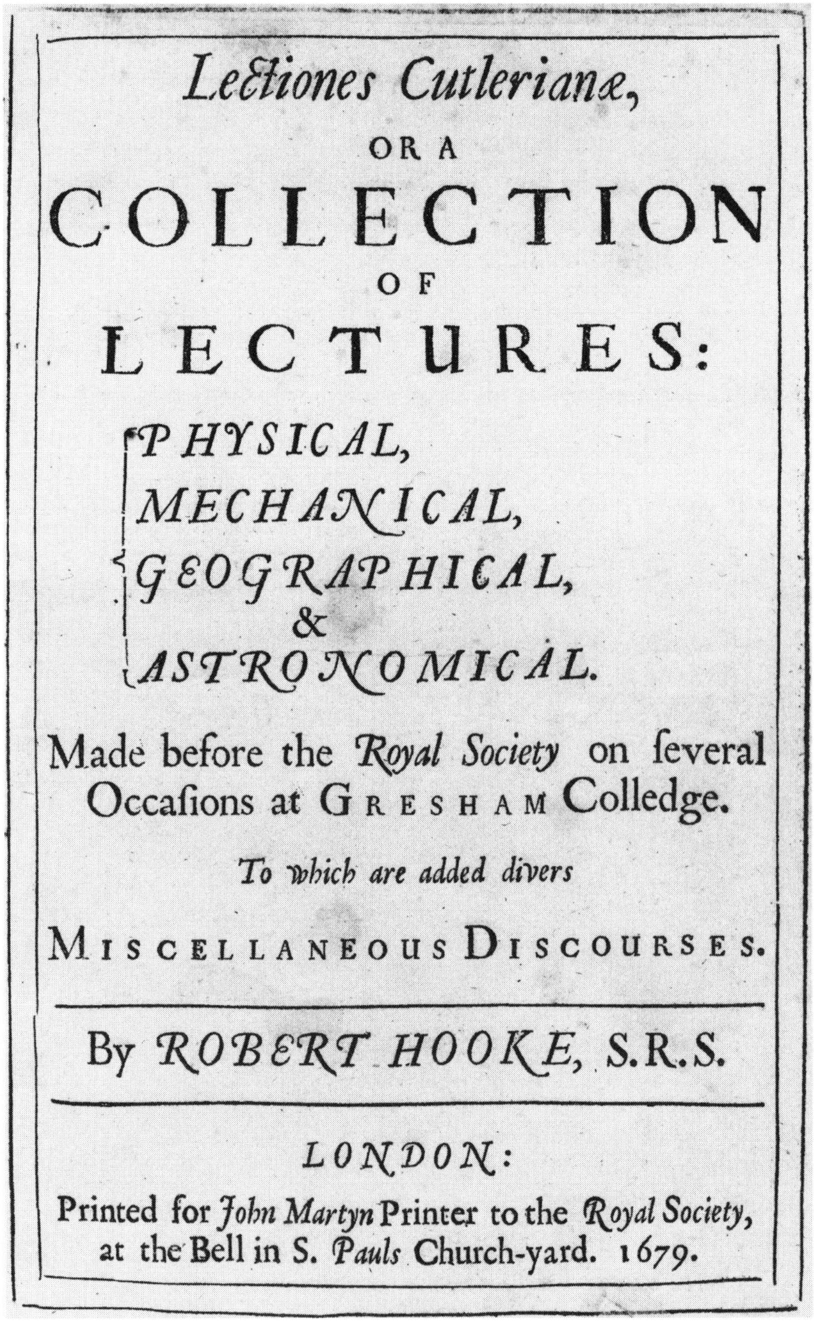

Lectiones Cutlerianæ,
OR A
COLLECTION
OF
LECTURES:
PHYSICAL,
MECHANICAL,
GEOGRAPHICAL,
&
ASTRONOMICAL.
Made before the *Royal Society* on ſeveral Occaſions at GRESHAM Colledge.
To which are added divers
MISCELLANEOUS DISCOURSES.

By *ROBERT HOOKE*, S.R.S.

LONDON:
Printed for *John Martyn* Printer to the *Royal Society*, at the Bell in S. *Pauls* Church-yard. 1679.

Fig. 36 The title-page of the collection of *Lectiones Cutlerianæ* issued by Hooke in 1679. (Museum of the History of Science, Oxford.)

stress that, without the Royal Society stimulus, it is doubtful if Hooke's *magnum opus* would have got into print at all.

Of Hooke's subsequent publications, the *Philosophical Collections* that he brought out between 1679 and 1682 constituted one of his responsibilities as Secretary to the Society in those years, and even his published

Cutlerian Lectures can be seen in a similar light. There seems little doubt that the publication of these formed part of Hooke's campaign to force Cutler to pay the arrears of his salary which started to accumulate in the 1670s, and the triumphal reissue of the series under the title of *Lectiones Cutlerianae* in 1679 did indeed fall at a critical point in the negotiations between the two men (Fig. 36).[6] Once he had established his legal entitlement to his arrears through his Chancery suit against Cutler—in 1683–4—Hooke again became nonchalant about publication, and no further works materialized during his lifetime, it being left to Richard Waller and William Derham to bring out posthumous collections of the manuscript material that Hooke left behind him in the form of the *Posthumous Works* of 1705 (Fig. 37) and *Philosophical Experiments and Observations* of 1726.[7]

Turning to the subject-matter of these writings, there are two components. As far as the Royal Society was concerned, what Hooke was employed to provide were 'demonstrations' at meetings. These could take various forms, and a number of them have already figured in Jim Bennett's section, since instruments and their demonstration and trial were always central. Hooke is frequently to be found demonstrating a piece of equipment and its potential, while he might also be charged with investigating topics suggested by other Fellows. But clearly part of the reason for his being employed as Curator of Experiments from November 1662 onwards was to try to escape from the kind of miscellaneousness in its proceedings to which the Royal Society had been prone almost from the outset, and Hooke was clearly expected to develop a systematic programme of investigation on specific themes. Initially, the topic chosen was pneumatics, on which Hooke carried out a rigorous series of experiments which unfolded from the end of 1662 onwards, on 25 February 1663 producing 'a scheme of inquiries concerning the air' which he then pursued over the following months: this represented a sequel to the experiments that Boyle had by this time published in his *New Experiments Physico-Mechanical, Touching the Spring of the Air and its Effects* (1660), which ran in parallel with further work by Boyle himself.[8] Then, it was microscopy: on 25 March, Hooke was solicited to prosecute his microscopical observations, being requested at the next meeting to bring in at least one microscopical observation each week, and so he did for nearly a year.[9]

Subsequently, in 1666, in the aftermath of the adjournment caused by the plague, a comparable programme of research was to unfold on the nature of attraction and its action on bodies, especially its role in celestial mechanics, which is again indicative of the way in which Hooke's commitment to the Royal Society encouraged him to develop

The Poſthumous

WORKS

OF

ROBERT HOOKE, *M.D. S.R.S.*

Geom. Prof. Greſh. *&c.*

Containing his

Cutlerian Lectures,

AND OTHER

DISCOURSES,

Read at the MEETINGS of the Illuſtrious

ROYAL SOCIETY.

IN WHICH

I. The preſent Deficiency of NATURAL PHILOSOPHY is diſcourſed of, with the Methods of rendering it more certain and beneficial.

II. The Nature, Motion and Effects of LIGHT are treated of, particularly that of the *Sun* and *Comets.*

III. An Hypothetical Explication of MEMORY; how the Organs made uſe of by the Mind in its Operation may be Mechanically underſtood.

IV. An Hypotheſis and Explication of the cauſe of GRAVITY, or GRAVITATION, MAGNETISM, *&c.*

V. Diſcourſes of EARTHQUAKES, their *Cauſes* and *Effects*, and Hiſtories of ſeveral; to which are annext, *Phyſical Explications* of ſeveral of the Fables in *Ovid's Metamorphoſes*, very different from other Mythologick Interpreters.

VI. Lectures for improving NAVIGATION and ASTRONOMY, with the Deſcriptions of ſeveral new and uſeful *Inſtruments* and *Contrivances*; the whole full of curious Diſquiſitions and Experiments.

Illuſtrated with SCULPTURES.

To theſe DISCOURSES is prefixt the AUTHOR'S LIFE, giving an Account of his Studies and Employments, with an Enumeration of the many Experiments, Inſtruments, Contrivances and Inventions, by him made and produc'd as Curator of Experiments to the *Royal Society*.

PUBLISH'D

By *RICHARD WALLER*, R. S. Secr.

LONDON:

Printed by SAM. SMITH and BENJ. WALFORD, (Printers to the Royal Society) at the *Princes Arms* in St. *Paul's* Church-yard. 1705.

Fig. 37 The title-page of the edition of Hooke's *Posthumous Works* brought out by Richard Waller in 1705. (Museum of the History of Science, Oxford.)

an experimental programme. In part, this was concerned with magnetism, involving an attempt to measure the force of attraction at increasing distances from the magnet. In addition, having noted that the motion of the celestial bodies might be represented by pendulums, Hooke presented a paper 'concerning the inflection of a direct motion into a curve by a supervening attractive principle'. This has been seen as one of the most critical contributions to an understanding of the problem of celestial mechanics to be produced in this period, and it illustrates Hooke's ability to apply principles from one field to another which Jim Bennett has already indicated as one of his great strengths, in that 'inflection' was initially an optical concept which he applied to cosmology.[10]

Many of the demonstrations that Hooke made are known only from the brief summary of them given in the Society's minutes. But in other cases Hooke was requested to submit a written account which survives in the Register Book or the Classified Papers—a case in point is the paper on inflection just referred to—and a number of these were published in William Derham's *Philosophical Experiments and Observations* or in Thomas Birch's *History of the Royal Society*.

With lectures as against demonstrations, the position is slightly more complicated. The model was evidently the Gresham Lectures, the series of public lectures delivered twice weekly during the law terms by the professors of Gresham College. These lectures had a fairly conservative rubric, but were clearly intended to give an introduction to the topics covered by the lecturer's terms of reference—in Hooke's case, geometry. But it seems to have been up to the lecturer to decide exactly what he wanted to do. Thus in the case of the astronomy lectures which John Flamsteed gave, deputizing for Walter Pope, the rubric was conservative, but the text that survives of the lectures actually given shows that they ranged widely and innovatively.[11] The same is true of Hooke's surviving geometry lectures, which included a good deal of what we would call natural philosophy, and hence form part of the corpus that I will be addressing here.

As for the Cutlerian Lectures, here matters are more complicated still. The basic idea was presumably for a lecture series comparable to the Gresham Lectures, which they were evidently seen as complementing in terms of their incidence, in that they took place in the vacations whereas the Gresham Lectures took place in the term.[12] They were presumably also intended to match the Gresham Lectures in giving an exposition of the principles of a subject. But what was the subject? This had been a matter of dispute, if not of sleight of hand, when prominent Fellows of the Royal Society negotiated with Cutler over the terms of his endowment. It seems fairly clear that Cutler, perhaps as befitted a merchant, wanted them to be about the history of trades, in other words,

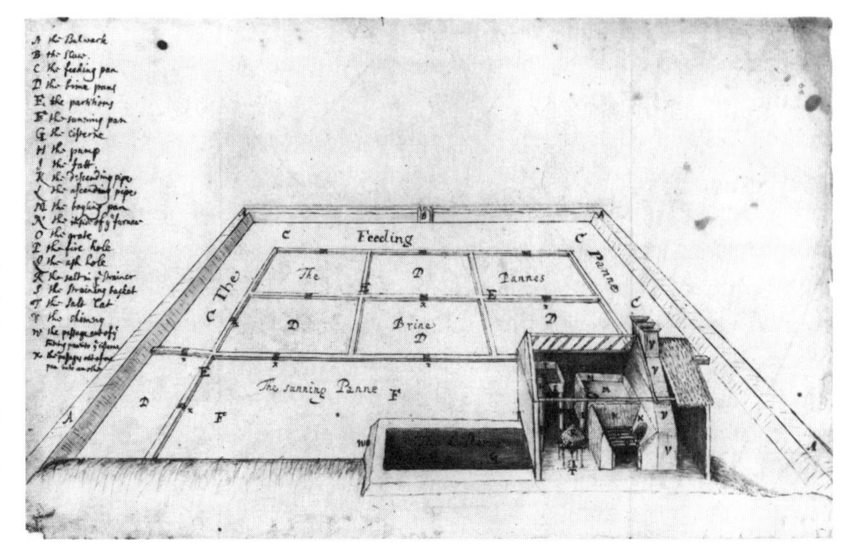

Fig. 38 Ink and grey wash drawing by Hooke illustrating 'the manner of making Salt at a Saltern in Hampshire', evidently relating to one of his early Cutlerian Lectures. (Royal Society Classified Papers XX, no. 40.)

the description of craft practices which was central to the early ambitions of the Royal Society, and which it was thought might result in improvements to the techniques involved. However, the Royal Society managed to substitute a somewhat different wording, which Cutler may have thought was a synonym for this, but which was actually susceptible to a quite different interpretation, namely 'the *History of Nature and Art*'.[13]

In a sense, the history of the Cutlerian Lectures represents an increasing movement away from Cutler's original intentions and towards topics which evidently seemed more interesting to the majority of Fellows of the Royal Society, not to mention Hooke himself. Initially, it seems that Hooke did actually give a couple of lectures on the history of trades. These survive in manuscript at the Royal Society, one dealing with salt-making in Hampshire (no doubt familiar to Hooke from his childhood in the Isle of Wight) (Fig. 38), the other dealing with the felt-maker's trade.[14] But Hooke soon seems to have tired of this, and his subsequent lectures interpreted 'the history of nature and art' in rather different ways. Indeed, this proved to be quite a brilliant formula, enabling Hooke to discuss a very broad range of aspects of the natural world, its study and exploitation.

Thus the Cutlerian Lectures published in the 1670s seem largely to have interpreted the rubric as referring to the way in which the construction of instruments, through art, would enhance our accurate understanding of nature. On the other hand, the other topic that Hooke seems to have felt appropriate comprised general methodological strictures on the proper procedures for conducting scientific research. These start in the 1660s: indeed, the germ of Hooke's seminal work, *A*

General Scheme, or Idea of the Present State of Natural Philosophy, can be traced to his lectures on the Cutlerian foundation. But this continued later, and comparable methodological prescriptions are also disproportionately prominent in his later Cutlerian Lectures. On the other hand, one of the great virtues of the 'art and nature' formula was that, under it, almost anything could be introduced as an exemplification of the subject-matter and method of science. Hooke was thus given a fairly free hand in his choice of subject. One topic to which he devoted extensive attention in his later lectures was geology. Another was the value of observations from distant lands, in terms both of knowledge and skills which one culture might have while another lacked, and of crucial data through which hypotheses might be tested.[15]

There is a change between these later Cutlerian Lectures and the early ones, in that, whereas the earlier ones had been regarded as separate from the business of the Royal Society (though held on the same day, an hour before the Society's meetings), now they were invariably seen as part of it, and hence they were minuted.[16] This no doubt reflects the problems that the Society encountered after the first flush of enthusiasm had worn off in keeping its meetings supplied with 'entertainment', but it paradoxically helps in keeping track of this aspect of Hooke's activities, which had formerly been badly documented.

The Theorist of Science

Let us begin by considering the text by Hooke that seems to be linked most closely with his early Cutlerian Lectures, his *General Scheme*, his most important methodological writing on science, which comprises an exposition of the proper method of compiling a natural history. Initially, this seems to have taken the form of a group of Cutlerian Lectures (the Cutlerian Lectures were delivered in groups, which made for a sustained exposition of a topic). Two entire lectures from this series, and a fragment of another, have survived, of which one complete one and the fragment have been published in full, while the other has been discussed at length.[17] These overlap substantially with the written-up version which Hooke then prepared, perhaps for publication, though in the event this was not to materialize till after his death in Waller's edition of the *Posthumous Works*. The title of the published version is: *A General Scheme, or Idea of the Present State of Natural Philosophy, And How its Defects may be Remedied By a Methodological Proceeding in the making Experiments and collecting Observations. Whereby to Compile a Natural History, as the Solid Basis for the Superstructure of True Philosophy* (Fig. 39).

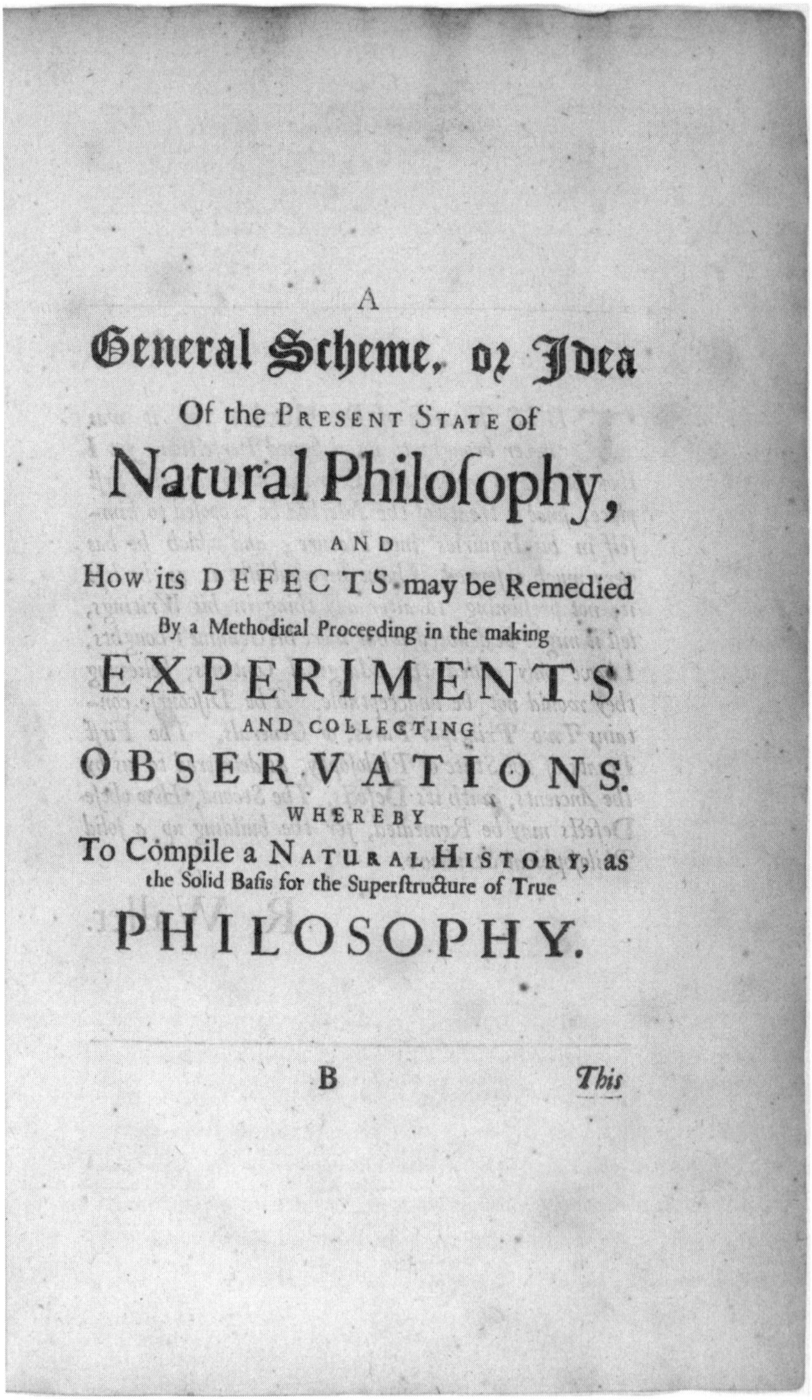

A

General Scheme, or Idea

Of the Present State of

Natural Philoſophy,

AND

How its DEFECTS may be Remedied

By a Methodical Proceeding in the making

EXPERIMENTS

AND COLLECTING

OBSERVATIONS.

WHEREBY

To Compile a Natural History, as the Solid Baſis for the Superſtructure of True

PHILOSOPHY.

B *This*

Fig. 39 The title-page of Hooke's 'General Scheme' as published in Waller's edition of his *Posthumous Works* (1705), p. 1. (Museum of the History of Science, Oxford.)

In his aspiration to a 'natural history' as the basis for a true philosophy, Hooke here echoes Francis Bacon, the early Stuart statesman and philosopher, who had outlined a programme for the reform of knowledge in a series of writings which he grouped together under the title, *The Great Instauration*.[18] Basically, Bacon argued that existing ideas about the workings of nature suffered from a premature pursuit of generalization on the basis of inadequate data, which had the effect of leading to 'disputes and scrappy controversies': he thus gave expression to the dissatisfaction with the science of the 'schools', of the universities dominated by the ideas of the ancient Greek philosopher, Aristotle, which was widespread in his period.

Bacon's remedy—expounded most fully in his methodological treatise, the *New Organon*, published in 1620—was to replace the scholastic use of deductive, syllogistic reasoning by the use of induction. Moreover, he insisted that the first essential was to build up a bank of accurate data about the natural world, and only then to attempt to reach general conclusions. For this purpose, he advocated the compilation of a comprehensive body of what he called 'natural histories', collections of rigorously verified empirical data about the natural world. It was this programme that Hooke presented in a reinvigorated form for his Restoration audience. (In fact, he rarely mentions Bacon by name, but I think that this was because he could take a familiarity with Bacon for granted rather than because he was claiming any inappropriate originality.) Indeed, the very concept of the Cutlerian foundation as being devoted to 'the history of nature and art' has clear Baconian roots, in that Bacon was particularly interested in the interrelationship between the two, and particularly the extent to which, through human manipulation, significant data about nature that would otherwise be overlooked could be brought to light. Like Bacon, Hooke was critical of the legacy of Aristotle and other ancients and the extent to which people were prone to making premature generalizations; like Bacon, he saw the key to the progress of knowledge as the accumulation of accurate data, on the basis of which reliable generalizations could be reached; and, like Bacon, he saw the answer in 'collecting a Philosophical History, which shall be as the Repository of Materials, out of which a new and sound Body of Philosophy may be raised'. Moreover, within this, he attached special significance to what could be learnt about nature, not just through passive observation, but by experimental manipulation.[19]

How were we to be sure that the information that we had was reliable? Both Bacon and Hooke saw it as crucial to be on our guard against standard sources of misapprehension and falsehood. In Bacon's case, this was presented in his famous metaphor of the four 'idols' of the human mind which needed to be eradicated. These were the idols of the cave—the effects of individual human temperament and upbringing; of the tribe—the effects of human nature in general; of the market—the tyranny of words and clichés; and of the theatre—the elaborate shows contrived by builders of philosophical 'systems' of the natural world. Hooke's diagnosis was similar, though differently organized: he started by indicating the defects of the human senses, evidently intended 'for some other Use than for the acquiring of this kind of Knowledge'; secondly, he indicated the extent to which individual personalities had an undue effect on people's studies and the conclusions they drew in them; finally, he traced prejudice to 'Language, Education, Breeding, Conversation, Instruction, Study, from an Esteem of Authors, Tutors, Masters, Antiquity, Novelty, Fashions, Customs, or the like'.[20]

Both men saw the need to be on guard against these sources of misapprehension about nature, but where Hooke went far beyond Bacon was in his insistence that many of the shortcomings of the senses could be rectified by the use of instruments, as Jim Bennett has shown. Thus, at this point in his *General Scheme* he goes into detail about the defects of the eye and the ways in which these could be mitigated by the use of optical instruments, while he also instanced an instrument for measuring the velocity of falling bodies that he had demonstrated to the Royal Society in 1663.[21] Subsequently, he went into detail about the ways in which such gadgets could be used to provide accurate data that would elude the unassisted senses.

Both men also insisted that it was imperative that data should be recorded as systematically as possible. In Bacon's case, he laid out a complete survey of the topics on which he thought natural histories should be compiled, his *Parasceve, or Preparative to a Natural and Experimental History*, while in the second part of his *New Organon* he showed in detail how he believed general truths about a particular natural phenomenon would emerge from a systematic compilation of relevant data in 'tables' and 'a co-ordination of instances'. In his *General Scheme,* Hooke again took up this ambition and developed it. Thus he started with an elaborate tabulation of data about natural phenomena, ranging from a very general classification of the visible world, both terrestrial and celestial, and then each component within it. He then drew up a list of 'Histories' that needed to be compiled—echoing Bacon's *Parasceve*, but again developing it in different ways, perhaps not least in his very lengthy classified list of craft techniques. Lastly, he drew up sample lists of queries concerning specific topics—the nature of comets, stars, and planets, the aether and atmosphere, and other aspects of the air.[22] Again, there were Baconian precedents, in terms of the queries that Bacon had prefixed to each of the 'Histories' that he had seen as exemplifying the method of his *Great Instauration*. But Hooke went beyond this, particularly in the implication that the lists that he provided in the *General Scheme* were merely a sample of such lists that it would be an easy matter to produce for all the other phenomena of the universe. Indeed, at least in potential, he went beyond his mentor, Boyle, who followed Bacon's example in producing a more limited range of lists of 'heads' of distinct topics.[23] In this regard, Hooke has claims to be considered one of the most systematic exponents of scientific method of his age.

As this indicates, it was axiomatic for Hooke—as for Bacon—that the pursuit of knowledge would be necessarily cooperative, involving the activity of a whole group of individuals working together. Moreover, in Hooke's case the obvious vehicle for this was the Royal Society, and it is therefore significant that the subject matter of the *General Scheme* seems originally to have been divulged as Cutlerian Lectures: it would thus have been addressed to the Fellows of the Royal Society who might, cumulatively, implement the immense programme of data collecting that Hooke put forward in it. He also exemplified the potential for collaborative enquiry, for instance, in the scheme for compiling a history of the weather that was published in Thomas Sprat's *History of the Royal Society* (1667), while the issue of how best to organize the collective efforts of the Society's members were to preoccupy him in the 1670s, when he drew up detailed proposals for the reform of that body to make it more effective in this regard.[24] It is also interesting that Hooke's later Cutlerian Lectures lay repeated stress on the aims and achievements appropriate to a corporate body in terms of collecting and systematizing data, in contrast to the efforts of individuals: he is constantly harping on about what could be achieved by such a corporation, in contrast to the efforts of 'private men'.[25] Moreover, especially in the years around 1680, after the death of the Society's supreme 'intelligencer', Henry Oldenburg, and before younger men came to the fore, Hooke was to play a significant role in collecting and disseminating data through the publication of his *Lectures and Collections* and *Philosophical Collections* (a temporary replacement for Oldenburg's highly successful *Philosophical Transactions*). Indeed, it could be argued that, if the Royal Society saw itself as embodying Bacon's utopian vision of 'Solomon's House', it was Hooke who did more than anyone to try to bring that vision to fruition.

In the *General Scheme*, Hooke went into detail about 'the Requisites in a Natural Historian', which in some respects recapitulated his prescriptions for data collecting elsewhere in that work—the ideal historian would be resistant to prejudice, for instance, and would record things systematically—but he also made it clear that such a figure should be alert to hypotheses, indeed, that he should have these in mind as he made his enquiries. As he put it: 'by this Means the Mind will be somewhat more ready at guessing at the Solution of many Phenomena almost at first Sight, and thereby be much more prompt at making Queries, and at tracing the Subtilty of Nature, and in discovering and searching into the true Reason of things.'[26] For all Hooke's stress on the importance of compiling accurate data in the form of a natural history, he was not one of those naive Baconians in the early Royal Society who disavowed speculation 'before the *Histories* of *Art* and *Nature* are compleatly done'.[27] It is important to stress that these were not all well-intended but mediocre compilers of miscellaneous natural histories like such chorographers as Joshua Childrey or Robert Plot, since it is now clear that no less a figure than the philosopher, John Locke, took a strongly antihypotheticalist line in natural philosophy, believing that the compilation of descriptive natural histories was the highest appropriate goal and that speculation was out of order.[28]

Yet, in this, Hooke again has a good deal in common with Bacon, who similarly believed—in contrast to the more naive of his followers—that 'axioms' should be tested against data as it was accumulated and vice versa. But Hooke arguably took this approach further than his progenitor, developing a much more sophisticated analytical method.

Indeed, here we come to the most intriguing—and difficult—aspect of Hooke's philosophy of science, and that is the mutual relationship between data-gathering and the formulation of 'axioms and theories', and the method by which he believed that it was proper to progress from one to the other. Early in the text that I have been expounding, Hooke speaks of his proposals for what 'may not improperly be call'd a Philosophical Algebra, or an Art of directing the Mind in the search after Philosophical Truths'; this has been much discussed ever since the time of Hooke's posthumous editor, Richard Waller, who speculated that Hooke might have planned but failed to implement—or at least failed to leave to posterity—an explicit form of mathematically based deduction.[29] More recently, by far the most sustained attack on this question was published by Mary Hesse in an article in *Isis* in 1966, in which she canvassed Waller's and other points of view in the context of a full appraisal of Hooke's views and of twentieth-century theories of 'the logic of scientific discovery'; subsequently, the issue has been addressed by David Oldroyd and Patri Pugliese.[30] What can we conclude?

The possibility favoured by Waller was that Hooke was invoking a truly mathematical model of deduction. Hence, at the end of Hooke's *General Scheme*, Waller attempted to plug the gap by appending some extracts from Hooke's Gresham Geometry Lectures to illustrate his belief that 'from a few self evident Axiomes and Definitions, and *Postulata* easy to be granted, a vast Structure of undeniable Truths have been raised.'[31] In this scenario, Hooke could perhaps be seen as invoking the model of the ancient mathematician, Euclid, whose *Elements* we know he had interpreted to Boyle when working for him in the late 1650s, and it is interesting that in *Micrographia* he was to invoke the '*Elements of Geometry*' as a model for his analytical processes.[32] Yet it is important to distinguish between an analogy with geometry as a model for clear ratiocination and the actual use of specifically geometrical techniques; whereas nothing that Hooke says suggests a direct mathematical analogy of the latter kind, there are quite strong clues suggesting the former.

An alternative is that the 'philosophical algebra' was linked to the ambitions that were widespread in the seventeenth century to devise a new, 'philosophical', language, which would have a clarity and logic lacking from existing tongues. This was again an interest to be found in the prescient Bacon, but it was especially associated with two of Hooke's principal mentors at Oxford in the 1650s, John Wilkins and Seth Ward, and hence it would make a lot of sense for Hooke to invoke it. It is worth noting that he was closely involved in the discussions that went into Wilkins's *Essay Towards a Real Character, and a Philosophical Language* of 1668, and in the attempts to revise it after his death. In addition, an interesting exchange of letters on this subject in 1681 with the German natural philosopher, G. W. Leibniz, survives, in which, commenting on the rationale of Wilkins's *Essay*, Hooke wrote: 'my aymes have Always been much higher, ‹vizt.› to make it not only usefull for Expressing & Remembring of things and notions but to Direct Regulate assist and even ‹necesitate &› compell the mind to find out and comprehend whatsoever is knowable.'[33] Yet, if Hooke did indeed conceive the potential of such a language in terms of philosophical analysis, rather than simply communication, this too was an idea that he left frustratingly undeveloped.

In any case, the third, and in my view the most likely, possibility is that advocated in slightly different ways by Mary Hesse and David Oldroyd, namely that Hooke's 'philosophical algebra' simply involved the systematic tabulation of alternative methods of explaining a phenomenon, or approaching a problem, which would make it easier to eliminate alternatives until only one correct answer remained. This could be expressed in tabular form, and there are a number of such documents among Hooke's papers, while in other cases Hooke's ratiocination can comparably be reduced to a table, as both Hesse and Oldroyd have shown.[34] The rationale was apparently set out most clearly by Hooke in a later document, in which he wrote:

> I call that true *Ratiocination* from such Sense, where being sure of the Premises, the Conclusion necessarily follows from them; which is the method of Reasoning made use of in *Geometry*, and by which we arrive at as great a certainty of things unseen as seen . . . Now tho' in Physical Inquiries, by reason of the abstruseness of Causes, and the limited Power of the Senses we cannot thus reason, and without many Inductions from a multitude of Particulars come to raise exact Definitions of things and general Propositions; yet by comparing the varieties of such Inductions we may arrive to so great an assurance and limitation of Propositions as will at least be sufficient to ground Conjectures upon, which may serve for making *Hypotheses* fit to be enquired into by the *Analytick* method, and thence to find out what other Experiments or Observations are necessary to be procured for further progress in the *Synthetick*.[35]

All this may seem a little inconclusive, and, in the absence of Hooke's actual exposition of his views on these topics, it is likely to remain so. But what is significant about it is its illustration that Hooke saw the construction of general theories and the role of hypotheses as central to science. This is also strongly exemplified by Hooke's most famous book, *Micrographia*, to which I now wish to turn.

Micrographia

As I have already explained, *Micrographia* was in many ways perceived as a Royal Society venture. We have already seen how Hooke was encouraged to make a sustained series of presentations on such topics at meetings of the Society in 1663, while at one point in the published text he specifically notes that the account of petrifaction that he includes was made by appointment of the Royal Society. It is also worth noting that the findings of other illustrious Fellows of the Society are repeatedly cited.[36] The idea of the observations being published was there from the outset: at the meeting on 25 March 1663, Hooke 'was solicited to prosecute his microscropical observations, in order to publish them'. The issue of printing arose again in March 1664 and it was thereupon arranged for his manuscript to be perused by Fellows of the Royal Society, starting with the President, Lord Brouncker. Finally, in November, the Society's imprimatur was granted (Fig. 40).[37]

This sense of the work as a 'Royal Society' book was reinforced by its presentation. Thus the fact that Hooke was an FRS was stated both on the title-page and in the official (Royal Society) imprimatur that faced it, while the title-page prominently bore the Society's coat of arms. After the Epistle Dedicatory to the King, Hooke included a second dedication 'To the Royal Society', while he also devoted a significant part of his preface to a grandiloquent account of the Society's aims and achievements, the rest presenting a shorter exposition of some of the themes of his *General Scheme* (to which he evidently there alludes), including the need for a new approach to knowledge and the role that instruments could play in this. Both here and throughout the work he stresses the need 'to begin to build anew upon a sure Foundation of Experiments', while he is also highly complimentary about Bacon, 'the thrice Noble and Learned *Verulam*'.[38]

The book did indeed show off the work that Hooke had done at the Royal Society's behest. Apart from anything else, it featured instruments, as Jim Bennett has explained: even within the preface, various instruments are described, and others are referred to thereafter, such as thermometers, hygroscopes, or telescopes.[39] And of course, among instruments, the book's whole *raison d'être* was to illustrate the findings of the microscope. Indeed, to some extent Hooke restricted its subject-matter in these terms, as where he disavowed discussion of the art of dye-stuffs, 'of which particulars, because our *Microscope* affords us very little information, I shall add nothing more at present'.[40]

Now clearly the background to this is the Royal Society's sense of defensiveness in its early years, its need to prove itself against those who said, 'What have they done?' By this time, an elaborate defence of the Society in the form of Thomas Sprat's *History* had been begun, but in fact its progress had slowed to a snail's pace, and it was far from finished by the time of the Great Plague in 1665, being finally published only in 1667.[41] Hence the publication of the findings of the Society's Curator of Experiments might well have seemed a perfect way of vindicating the Society.

For these purposes, the ostensible subject-matter of *Micrographia* was ideal—namely its detailed and surprising revelations about the intricacy of what might be called the micro-world that surrounds us. Indeed, this is precisely what Hooke made a point of at the start of the book, illustrating how the point of a needle was really blunt, and how a 'point' in typography was a great smudge.[42] Thereafter, a good deal of the book, and the aspect of it that is most familiar to readers in retrospect, is its exposition of a series of microscopic observations illustrating the intricacy of nature.

Thus we have Hooke's famous examination of the composition of Kettering (in other words, Ketton) stone and other stones; of cork and mould; of the stinging nettle; and, perhaps above all, of various insects,

Fig. 40 The title-page of Hooke's *Micrographia* (1665), with the Royal Society's coat of arms, and, on the facing page, its imprimatur signed by Lord Brouncker as President. (Museum of the History of Science, Oxford.)

including the well-known large-scale plates of the flea and the louse (Fig. 41). In each case, Hooke lays out precise observations which arguably were the quintessence of the empiricism which the Royal Society championed. Let me give an example of this, because the success of the book

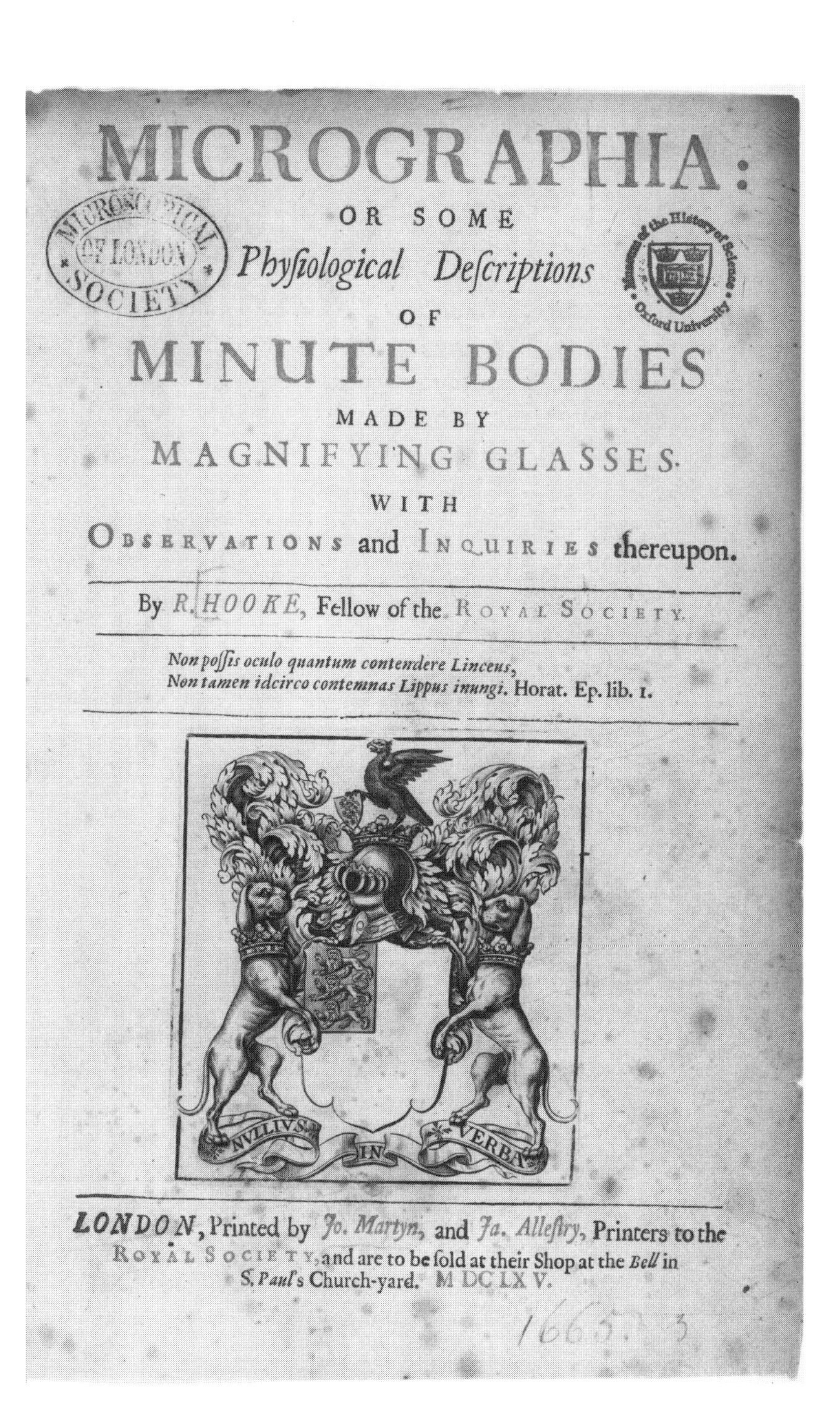

MICROGRAPHIA:
OR SOME
Physiological Descriptions
OF
MINUTE BODIES
MADE BY
MAGNIFYING GLASSES.
WITH
OBSERVATIONS and INQUIRIES thereupon.

By *R. HOOKE*, Fellow of the ROYAL SOCIETY.

Non possis oculo quantum contendere Linceus,
Non tamen idcirco contemnas Lippus inungi. Horat. Ep. lib. 1.

LONDON, Printed by *Jo. Martyn*, and *Ja. Allestry*, Printers to the ROYAL SOCIETY, and are to be sold at their Shop at the *Bell* in S. *Paul's* Church-yard. M DC LX V.

is undoubtedly partly explicable in terms of the tellingly accurate and full narratives that he provided. My choice is the stinging nettle (Fig. 42), where, having explained about the commonplace sensation of being stung by it, he continues:

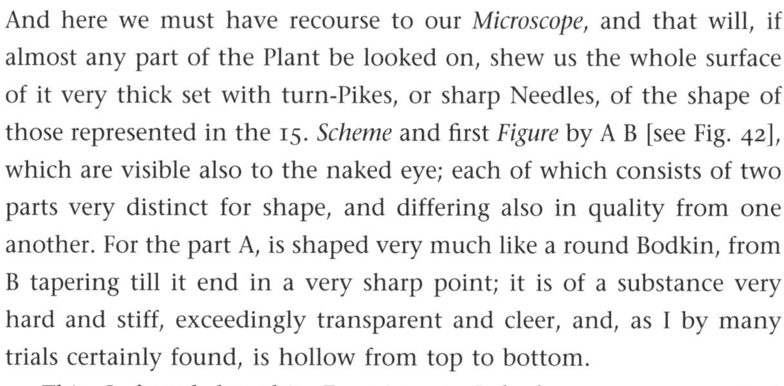

And here we must have recourse to our *Microscope*, and that will, if almost any part of the Plant be looked on, shew us the whole surface of it very thick set with turn-Pikes, or sharp Needles, of the shape of those represented in the 15. *Scheme* and first *Figure* by A B [see Fig. 42], which are visible also to the naked eye; each of which consists of two parts very distinct for shape, and differing also in quality from one another. For the part A, is shaped very much like a round Bodkin, from B tapering till it end in a very sharp point; it is of a substance very hard and stiff, exceedingly transparent and cleer, and, as I by many trials certainly found, is hollow from top to bottom.

This I found by this Experiment, I had a very convenient *Microscope* with a single Glass which drew about half an Inch [i.e., a microscope magnifying about twenty times], this I had fastned into a little frame, almost like a pair of Spectacles, which I placed before mine eyes, and so holding the leaf of a Nettle at a convenient distance from my eye, I did first, with the thrusting of several of these bristles into my skin, perceive that presently after I had thrust them in I felt the burning pain begin; next I observ'd in divers of them, that upon thrusting my finger against their tops, the Bodkin (if I may so call it) did not in the least bend, but I could perceive moving up and down within it a certain liquor, which upon thrusting the Bodkin against its basis, or bagg B, I could perceive to rise towards the top, and upon taking away my hand, I could see it again subside, and shrink into the bagg; this I did very often, and saw this *Phænomenon* as plain as I could ever see a parcel of water ascend and descend in a pipe of Glass. But the basis underneath these Bodkins on which they were fast, were made of a more pliable substance, and looked almost like a little bagg of green Leather, or rather resembled the shape and surface of a wilde Cucumber, or *cucumeris asinini,* and I could plainly perceive them to be certain little baggs, bladders, or receptacles full of water, or as I ghess, the liquor of the Plant, which was poisonous, and those small Bodkins were but the Syringe-pipes, or Glyster-pipes, which first made way into the skin, and then served to convey that poisonous juice, upon the pressing

Fig. 41 The engraving of the bluebottle and its open wing in *Micrographia*, a typical example of the exquisite studies of insects in that work (Schema XXVI). (Museum of the History of Science, Oxford.)

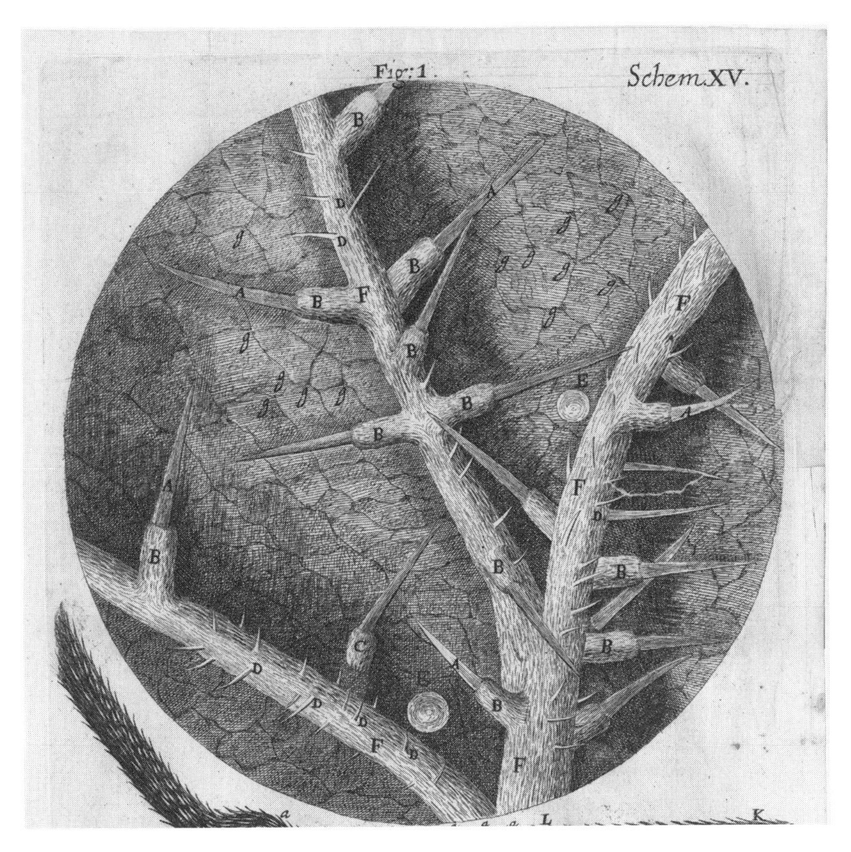

Fig. 42 The engraving of the stinging nettle in *Micrographia* (Schema XV). For Hooke's accompanying description, see pp. 127–8. (Museum of the History of Science, Oxford.)

> of those little baggs, into the interior and sensible parts of the skin, which being so discharg'd, does corrode, or, as it were, burn that part of the skin it touches; and this pain will sometimes last very long, according as the impression is made deeper or stronger.[43]

On the other hand, it is worth pointing out that the part of the book devoted to relatively straightforward microscopic observations really starts only about halfway through, whereas the early part of the book is dominated (apart from expositions of instruments) by a number of sustained passages in which Hooke set out theories about natural phenomena.

In particular, there is a long disquisition on congruity and incongruity (in fact a revised version of his previously published tract of 1661, a scarce item to the content of which he was doubtless keen to give wider circulation). Other topics that receive sustained treatment in this way include the nature of colour and light and of flame, and the process of petrifaction.[44] Such disquisitions also recur at the end of the book, one of them concerning what he described as 'inflection' in the air, and the other the nature of the moon and the conclusions that could be drawn from accurate observation of it.[45]

Indeed, for all its role as a manifesto for the careful empirical policy of the Royal Society, the book is far from restricted to a descriptive level. Hooke is constantly probing at *why* things are as they are. Thus he invoked 'the variety of reflection' as the reason 'why a small breez or gale of wind ruffling the surface of a smooth water, makes it appear black', offering as 'a very probable (at least, if not the true) cause of the hardning or tempering of Steel' the greater or lesser 'proportion of a vitrified Substance interspersed through the pores of the Steel'; he had similarly ingenious suggestions to make on all sorts of other topics.[46] Again and again he begins an excursus by explaining how 'The cause of all which *Phænomena* I imagine to be no other then this,' elsewhere using the phrase, 'this seems also to be the reason why', to introduce a further, subsidiary line of speculation.[47]

Some of these suggestions are fairly superficial, but others provide an elaborate process of ratiocination which brings us back to the issue of Hooke's method for moving from observation to theory, and hence the 'philosophical algebra' that I referred to earlier. Indeed here, as in his *General Scheme*, he aspires to 'some new engine and contrivance, some new kind of *Algebra*, or *Analytick Art*', interestingly citing Bacon in this connection, both in describing this as 'a *novum organum*', and in talking of a '*Scalam Intellectum*; he must have scaling Ladders, otherwise the steps are so large and high, there will be no getting up them.' In fact, in the thirteenth observation, he goes through an elaborate process of analysis which may be intended to exemplify his 'philosophical algebra', despite the fact that elsewhere his method seems casual.[48]

This raises some interesting questions as to what Hooke thought he was doing in the book, and what his Royal Society masters thought he was doing. It does seem as if the book caused the Society some unease. When the President was authorized to give it the Society's imprimatur, it was ordered

> That Mr HOOKE give notice in the dedication of that work to the society, that though they have licensed it, yet they own no theory, nor will be thought to do so: and that the several hypotheses and theories laid down by him therein, are not delivered as certainties, but as conjectures; and that he intends not at all to obtrude or expose them to the world as the opinion of the society.[49]

The background to *this* is provided by the extent to which, as we have seen, in addition to the sophisticated blend of natural history and hypotheticalism that Hooke exemplified, the Royal Society also contained a rather different strand of Baconianism, which insisted that theorization was premature until a complete bank of data had been collected, and the power of which within the Society's ranks can easily be underestimated.[50]

Quite what conclusions one should draw about this I am not sure. In view of what we know about Hooke's stance on such matters, and the constant fertility of speculation that he showed not only in *Micrographia* but in all his other writings, it would perhaps have been a little naive of the grandees of the Royal Society to try to rein him in. Indeed, they may not have intended to: the caveat about hypotheses almost reads like a kind of 'health warning' to a book they nevertheless found as stimulating as we do. (It also suggests that Hooke was less constrained by his status than some recent views of him have implied.)[51] For Hooke, on the other hand, it does seem as if this was an opportunity to make clear his speculations as much as his observations, thus getting himself taken seriously as a natural philosopher. And that did indeed occur, for *Micrographia* has been widely discussed almost ever since. Indeed, one of the most attentive of early readers was none other than Isaac Newton, whose notes 'out of Mr Hooks Micrographia' survive and have been published, showing how intrigued he was both by many of the phenomena that Hooke recorded and by Hooke's speculations about them.[52]

Hooke's Philosophy

It is useful to summarize here some of the main features of the philosophy of nature that Hooke puts forward. At the forefront was an ethos of clarity: 'Why should we endeavour to discover mysteries in that which has no such thing in it?'[53] The clarity that Hooke sought he found above all in the mechanical philosophy, the view that all phenomena could be explained in terms of the simple interaction of matter and motion, which was increasingly prevalent among natural philosophers of his day.

Fig. 43 René Descartes (1596–1650), whose ideas greatly influenced Hooke, but to whose non-empirical streak Hooke took exception. Engraved by I. Suÿderhoeff from a portrait by Frans Hals. (Museum of the History of Science, Oxford.)

In his rigorous mechanism, Hooke's chief inspiration was clearly the French philosopher, René Descartes (Fig. 43), who, more than anyone else, had provided the wherewithal for an alternative world view to Aristotle's through the fully mechanistic explication of the workings of the world that he expounded, notably in his *Principia philosophiæ* of 1644. Indeed, Descartes is the author to whom Hooke most frequently refers in *Micrographia*, and the *Principia philosophiæ* is the work that he cites most often. Many of his citations are of an implicitly approving nature, indicating how fully he had absorbed the view of the world that Descartes had put forward. He thus speaks of 'the ingenious *Des Cartes*' or 'the most acute and excellent Philosopher *Des Cartes*', frequently invoking Cartesian explanatory principles to account for phenomena.[54] This evidently reflects the exposure to Descartes's ideas that Hooke had had at Oxford during the 1650s.

It is easy to see the appeal of such views, for this was a system which was at the same time exhilaratingly simple, yet which offered endless potential for explaining phenomena in the world, and for their quantification. In his preface Hooke explained how he and his colleagues had come to believe that the effects of bodies that had previously been attributed to Aristotle's qualities 'are perform'd by the small *Machines* of Nature . . . seeming the meer products of *Motion, Figure*, and *Magnitude*'.[55] In his text, he was constantly seeking mechanical explanations of phenomena, taking it for granted 'that Nature does not onely work Mechanically', but that it could be elucidated by analogy to mechanical models—clocks, looms, ships, and the like—to make sense of the 'stupendious contrivances' by which it operated.[56] Qualities such as heat were to be explained in these terms: 'Nor need we suppose heat to be any thing else, besides such a motion,' he writes, while it is to mechanical principles that he repeatedly appeals in his explanations, for instance in the theory of light that he puts forward.[57]

Indeed, it was not least this view of the world having a mechanical micro-structure that the microscope could be used to substantiate. The detail exposed by the microscope showed that many things were invisibly small and could, therefore, be plausibly interpreted as being made of even smaller bodies. Its findings, in Hooke's words, revealed 'either *exceeding small Bodies*, or *exceeding small Pores*, or *exceeding small Motions*'.[58] The match between medium and message was ideal.

On the other hand, in *Micrographia*, Hooke shows himself critical of the brand of mechanical philosophy exemplified by Descartes in a way that is typical of English scientists in his period, finding it too hypothetical, content with plausible explanations which were never tested or assessed. As he explained in a crucial passage in Observation 8, concerning the sparks struck from a flint,

> we see by this Instance, how much Experiments may conduce to the regulating of *Philosophical notions*. For if the most Acute *Des Cartes* had applied himself experimentally to have examined what substance it was that caused that shining of the falling Sparks struck from a Flint and a Steel, he would certainly have a little altered his *Hypothesis*, and we should have found, that his Ingenious Principles would have admitted a very plausible Explication of this *Phænomenon*; whereas by not examining so far as he might, he has set down an Explication which Experiment do's contradict.[59]

In his recourse to 'experiment' as the ultimate authority, Hooke resembled his former employer, Robert Boyle (see Fig. 52), and this raises the question of the relationship between the two, which this is an appropriate place to explore. In fact, it seems possible that Hooke himself played a key role in introducing Boyle to Descartes in the first place, to judge from John Aubrey's comment in his *Brief Life* of Hooke, that he 'made him understand Des Cartes' Philosophy' (these words replace 'taught him', and the substitution is clearly intended to be significant); there is also evidence of Hooke reading lessons to Boyle concerning Descartes at a later date.[60] On the other hand, whether or not he had fully 'understood' Descartes till helped by Hooke, Boyle had been aware of the mechanical philosophy long before Hooke entered his employ. Like Hooke, he had early become a devotee of a mechanical view of nature for its clarity and intelligibility; but he, similarly, had become dissatisfied by the speculative nature of many of its explanatory claims, which he sought to test by increasingly systematic experimentation.

This began with chemical experiments—it was through chemistry that Boyle had been introduced to laboratory practice at the start of the 1650s—but by the end of that decade he was casting his net increasingly widely in attempting to vindicate his own variant of the mechanical philosophy, which he christened 'corpuscularianism' (a concept which Hooke clearly derived from him).[61] The exact date at which Hooke started to work for Boyle in the late 1650s is unclear, but the association between the two men was sealed by Hooke's ability to design and operate a piece of equipment, the vacuum chamber or air-pump, which allowed Boyle to fulfil a long-standing ambition to make an empirical investigation of the characteristics of the air. The trials in question took place in 1659, and they were published almost immediately in the form of the earliest and most celebrated of Boyle's scientific publications, his *New Experiments Physico-Mechanical, Touching the Spring of the Air and its Effects* (1660). In this work, Hooke is referred to only briefly, largely in connection with the construction of the pump (though Boyle does there state more generally that Hooke 'was with me when I had these things under consideration').[62] Despite the fact that this is all that Boyle directly states about their collaboration, it stands to reason that the two men must have had lengthy discussions of Boyle's findings and the conclusions that could be drawn from them about the nature of the air. Moreover, Hooke was to make a more substantial contribution to Boyle's subsequent *Defence* of his book in the form of a separate section which he authored, even though he was not publicly acknowledged for this as he should have been, due to a printer's error.[63] It is therefore appropriate that Hooke's earliest publication, which was subsequently expanded as one of the chief speculative sections of *Micrographia*, dealt with a topic arising from this book.

This was Hooke's theory of congruity, which stemmed from Boyle's demonstration of the phenomenon of capillary action in Experiment XXXV of *New Experiments*. Boyle there noted various phenomena concerning the abnormal behaviour of water in a thin glass tube, for which he offered no explanation, and it was this that Hooke furnished in terms of a notion of what he called 'congruity', which he defined as *'a property of a fluid Body, whereby any part of it is readily united with any other part, either of it self, or of any other Similar, fluid, or solid body'*; its converse, 'incongruity', was defined as *'a property of a fluid, by which it is hindred from uniting with any dissimilar, fluid, or solid Body'*.[64] Indeed, it was typical of the overtly speculative element in *Micrographia* that, whereas in his 1661 tract Hooke had refrained from offering a causal explanation of congruity (a nescience echoing that of Boyle himself concerning the 'spring' that he postulated in the air, of which it was sufficient to demonstrate the existence), now he openly speculated as to its explanation, linking it with the effects of motion on similar or dissimilar particles: he argued that whereas it made the former 'vibrate together in a kind of *Harmony* or *unison*', with the latter, 'though they may be both mov'd, yet are their *vibrations* so *different*, and so *untun'd*, as 'twere to each other, that they *cross* and *jar* against each other, and consequently, *cannot agree* together, but *fly back* from each other to their similar particles'.[65]

Later in *Micrographia*, Hooke offers a further digression which stems from the pneumatic work that he had done with Boyle, in this case setting out the findings concerning the relationship between the volume and pressure of air that the two men had studied together. (In parallel with this, the phenomenon had also been studied by Richard Towneley and Henry Power, but they did not draw out its implications as explicitly as Boyle did.) Boyle had initially propounded these findings in the *Defence* of his *New Experiments* that he published in 1662, and the formula involved is therefore called Boyle's Law: the experiment in question involved the celebrated J-tube, a long tube bent at the bottom to form two parallel sections, the shorter of which was sealed, which could be used to demonstrate that the volume of air is inversely proportional to the pressure on it. In his *Defence*, Boyle's primary aim was to refute the argument that had been put forward to counter his claims by the English Jesuit Aristotelian, Francis Linus. In contrast to Boyle's view that a vacuum could indeed by created by his new pump, Linus had argued that no vacuum existed, and that the phenomena observed in suction pumps and syphons were in fact to be attributed to a 'funiculus', a kind of internal thread. For the refutation of this view, more than basic quantification was not essential.[66] In *Micrographia*, however, Hooke was able to go much further, providing detailed tables illustrating the relationship involved, and developing the theory by using further experiments with the J-tube to consider the effect of altitude on air, with special reference to the uneven refraction of light (Fig. 44).[67]

Hooke's speculations about heat and combustion in *Micrographia* similarly represented the development of ideas shared with Boyle, in this case more overtly critical of the ideas of Descartes, who had explained combustion in terms of particles of fire breaking out of a heated body in the form of sparks and flames. Hooke was scornful of this idea, instead arguing that heat was '*a property of a body arising from the motion or agitation of its parts*': indeed, it was in this connection that he criticized Descartes for his insufficient attention to experiment in the passage that has already been cited. Moreover, developing this in another context later in the book, he speculated about the nature of combustion, which he argued resulted from a mixture of some part of the burning body with a component of the air.[68] The amount of this substance which the air contained seemed to be finite, so that combustion ceased without a fresh supply of air, and this led him to speculate further on the properties of air, and particularly about its role in respiration. This was an interest that he shared with Boyle and others who had worked with the two men at Oxford in the years around 1660, but which Hooke developed in an original and significant way which was taken up by other members of the Oxford group, such as John Mayow.

Indeed, here Hooke's ideas cross-fertilized mechanistic ideas with notions derived from the chemical tradition, ultimately stemming from the sixteenth-century iatrochemist, Paracelsus. This indicates the range of theories available at the time to those who were primarily empiricists, and who were dissatisfied with an unduly strict reading of the mechanical philosophy, not least in the circles at Oxford in which Hooke had moved in the 1650s.[69] Paracelsus had postulated that there might be a nitrous substance in the air and Hooke took up this view, arguing that air was 'nothing else but a kind of *tincture* or *solution* of terrestrial and aqueous particles', especially saline ones, and that it acted as a '*menstruum*, or universal dissolvent of all *Sulphureous* bodies' by virtue of 'a substance inherent, and mixt with the Air, that is like, if not the very same, with that which is fixt in *Salt-peter*'.[70]

There are also Hooke's speculations on light and colour, another topic on which, incidentally, his ideas resonated with Boyle's, particularly as published in his *Experiments and Considerations touching Colours* (1664), a book to which Hooke more than once defers.[71] In Hooke's case, his speculations stemmed from his observation of the iridescent colours seen

226 · MICROGRAPHIA.

A Table of the Elastick power of the Air, both Experimentally and Hypothetically calculated, according to its various Dimensions.

The dimensions of the included Air.	The height of the *Mercurial Cylinder* counterpois'd by the *Atmosphere.*	The *Mercurial Cylinder* added, or taken from the former.	The sum or difference of these two *Cylinders.*	What they ought to be according to the *Hypothesis.*
12	29 +	29 =	58	58
13	29 +	$24\frac{11}{16}$ =	$53\frac{11}{16}$	$53\frac{7}{13}$
14	29 +	$20\frac{3}{16}$ =	$49\frac{3}{16}$	$49\frac{5}{7}$
16	29 +	14 =	43	$43\frac{1}{2}$
18	29 +	$9\frac{1}{8}$ =	$38\frac{1}{8}$	$38\frac{2}{3}$
20	29 +	$5\frac{3}{16}$ =	$34\frac{3}{16}$	$34\frac{4}{5}$
24	29	0 =	29	29
48	29—	$14\frac{5}{8}$ =	$14\frac{3}{8}$	$14\frac{1}{2}$
96	29—	$22\frac{1}{8}$ =	$6\frac{7}{8}$	$7\frac{1}{4}$
192	20—	$25\frac{5}{8}$ =	$3\frac{3}{8}$	$3\frac{5}{8}$
384	29—	$27\frac{2}{8}$ =	$1\frac{6}{8}$	$1\frac{7}{16}$
576	29—	$27\frac{7}{8}$ =	$1\frac{1}{8}$	$1\frac{5}{24}$
768	29—	$28\frac{1}{8}$ =	$0\frac{7}{8}$	$0\frac{3}{4}$
960	29—	$28\frac{1}{8}$ =	$0\frac{7}{8}$	$0\frac{4}{5}$
1152	29—	$28\frac{7}{16}$ =	$0\frac{9}{16}$	$0\frac{10}{16}$

From

Fig. 44 Hooke's 'Table of the Elastick power of the Air', giving the results of his experiments using the J-tube, from *Micrographia*, p. 226. (Museum of the History of Science, Oxford.)

in thin plates such as mother of pearl, the succession of colours in which, he argued, was produced by the combination of light reflected from the upper and lower surfaces. From this, he went on to attack Descartes's view of how white light was converted into coloured light, arguing that when the pulses of light were refracted by the medium through which they travelled, they were converted into the primary colours blue and red, of which the other colours are 'dilutions'. His conclusion was 'That *Blue is an impression on the Retina of an oblique and confus'd pulse of light, whose weakest part precedes, and whose strongest follows*. And, that *Red is an impression on the Retina of an oblique and confus'd pulse of light, whose strongest part precedes, and whose weakest follows*.'[72]

Ranging still more widely, Hooke put forward seminal cosmological ideas, arguing that each primary planet and each satellite of a primary planet had its own gravitation, including the moon (Fig. 45). As he explained,

> the figure of the superficial parts of the Moon are so exactly shap'd, according as they should be, supposing it had a gravitating principle as the Earth has, that even the figure of those parts themselves is of sufficient efficacy to make the gravitation, and the other two suppositions probable . . . for I could never observe, among all the mountainous or prominent parts of the Moon (whereof there is a huge variety) that any one part of it was plac'd in such a manner, that if there should be a gravitating, or attracting principle in the body of the Moon, it would make that part to fall, or be mov'd out of its visible posture.

Again, this was a significant critique of Descartes's views, in this case his vortex theory of heavenly motion, and particularly his view that the moon was embedded in the vortex of the earth and must draw its gravitation from that. Indeed, as Hooke wrote in the penultimate sentence of the book, 'some other principle must be thought of, that will agree with all the secundary as well as the primary Planets.'[73] He thus intimated the enquiries that he was to pursue in subsequent years and which (as we shall see) were to lie behind his dispute with Newton.

The last of the examples of the 'speculative' side of *Micrographia* that I want to instance comprises Hooke's views on petrifaction. On the basis of his microscopic observations of them, in Observation 17 he argued against the view that was common at the time that fossils were 'lusus naturae', 'Stones form'd by some extraordinary *Plastick virtue latent* in the Earth it self'. Instead, he asserted that they were 'the Shells of certain Shelfishes, which, either by some Deluge, Inundation, Earthquake, or some such other means, came to be thrown to that place'.[74] Here is the germ of Hooke's later geological writings, to which we will return in due course.

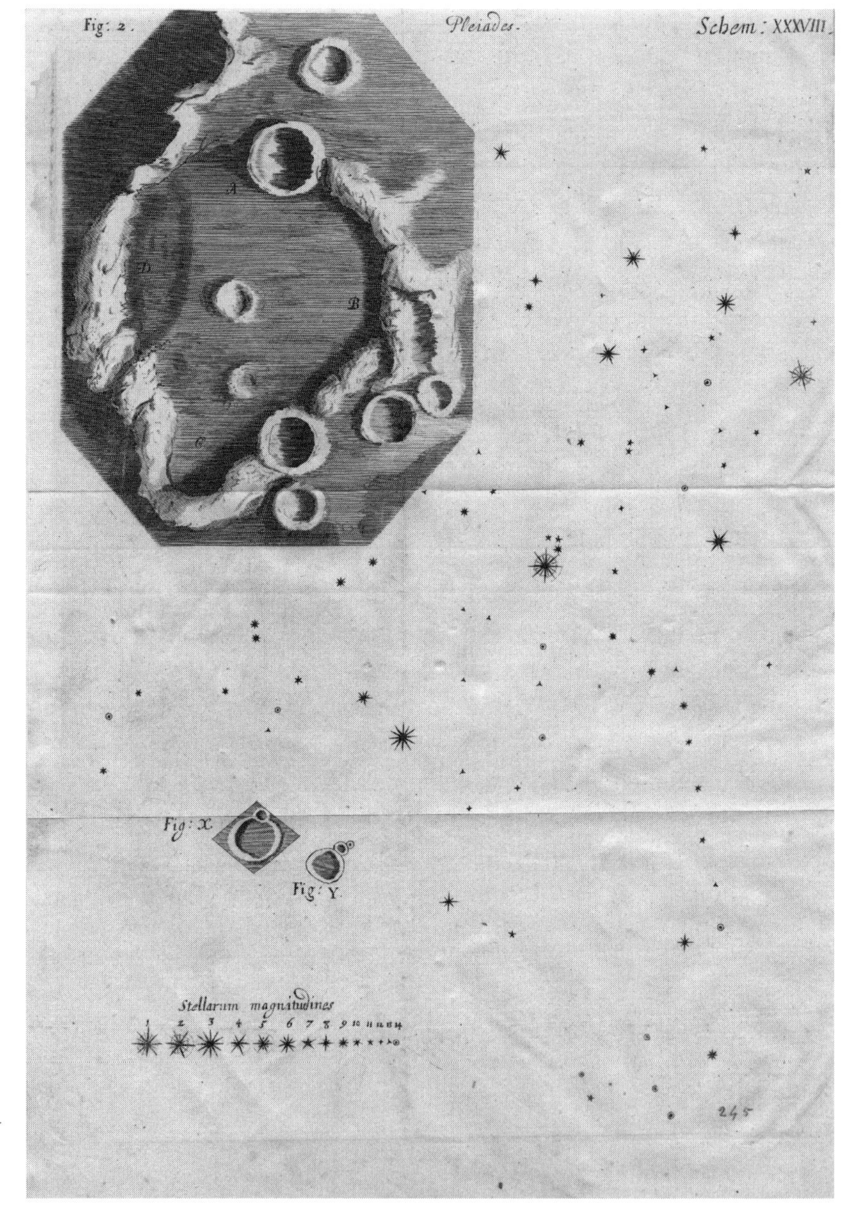

Fig. 45 Hooke's drawing of the surface of the moon, and of the 'multitudes of small Stars' discernible in the Pleiades by the use of a telescope, from *Micrographia*, Schema XXXVIII. (Museum of the History of Science, Oxford.)

It is perhaps worth noting that most of Hooke's speculations relate to the physical sciences, notwithstanding his view of the animate creation as being a higher species of being—'that seeming to be the highest step of natural knowledge that the mind of man is capable of'.[75] In the biological sciences, though he has been credited with enunciating the concept of the 'cell' for the first time, in fact he used it in a simple, descriptive sense, and he does not seem to have had a sense of the cell as the basic structural form of living things.[76] In fact, the biological sections of the book are characterized not so much by speculation as by assertion—of the proof that the findings of the microscope offer for the intricacy of nature and hence for the purposefulness of God in creating it. Repeatedly, Hooke includes such comments as that 'Nature, that knows best its own laws, and the several properties of bodies, knows also best how to adapt and fit them to her designed ends':

> And indeed, if we consider the great care of the Creator in the dispensations of his providences for the propagation and increase of the race, not onely of all kind of Animals, but even of Vegetables, we cannot chuse but admire and adore him for his Excellencies.[77]

To twenty-first-century readers, this can easily seem slightly banal. But I think that it is worth trying to look though seventeenth-century eyes, and to see *this* as a kind of grand speculation: in other words, that through painstaking analysis, we would understand aspects of God's intricate but often baffling handiwork that would otherwise have eluded us. In this, Hooke partakes of the tradition of natural theology that was so significant in the natural philosophy of the period.

Hooke's Later Writings

Having used *Micrographia* at such length to illustrate the character of Hooke's science, I will pass more briefly over his later writings. But this is because, to a large extent, these further exemplify the kind of themes to be found in *Micrographia*. Thus, although the Cutlerian Lectures of the 1670s are substantially devoted to instruments that Hooke had devised and their rationale, as Jim Bennett has already shown, Hooke at times draws out more general implications from the findings that he put forward in them. A classic case in point, epitomized even by its title, is his *Attempt to Prove the Motion of the Earth from Observations* (1674; Fig. 46), which ends with his notorious promise to 'explain a System of the World differing in many particulars from any yet known, answering in all things to the common Rules of Mechanical Motions'. This involved the supposition that all celestial bodies had a gravitational power which effected the motion of bodies that came into their orbit, and that the mutual relations of these could be calculated. He continued:

> Now what these several degrees are I have not yet experimentally verified; but it is a notion, which if fully prosecuted as it ought to be, will mightily assist the Astronomer to reduce all the Cœlestial Motions to a certain rule, which I doubt will never be done true without it.[78]

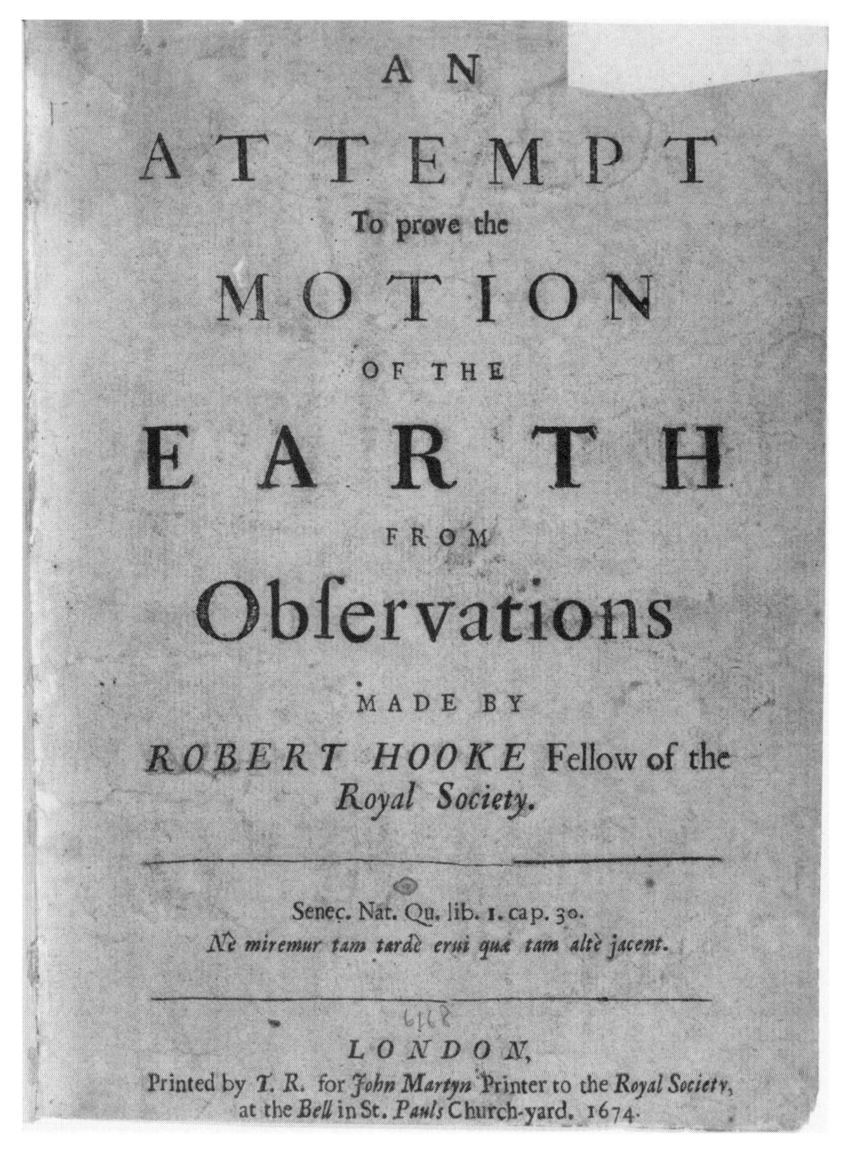

AN
ATTEMPT
To prove the
MOTION
OF THE
EARTH
FROM
Obſervations
MADE BY
ROBERT HOOKE Fellow of the
Royal Society.

Senec. Nat. Qu. lib. 1. cap. 30.
Ne miremur tam tardè erui quæ tam altè jacent.

LONDON,
Printed by *T. R.* for *John Martyn* Printer to the *Royal Society*, at the *Bell* in St. *Pauls* Church-yard. 1674.

Fig. 46 The title-page of Hooke's *Attempt to Prove the Motion of the Earth from Observations* (1674). (Museum of the History of Science, Oxford.)

It was this published comment which lay at the root of the subsequent feud between Hooke and Newton. Here, it is perhaps worth noting how Hooke's instincts led him to think in terms of experimental verification of the phenomenon, whereas Newton, of course, was to verify it mathematically.

In *Cometa, or Remarks about Comets* (1678), Hooke reverts to his typical methodological prescriptions, setting out 'certain Queries necessary to be answered, in order to find out the true theory of them' which he sought systematically to answer using observational data and speculations arising from it.[79] Then, in his *Lectures de Potentia Restitutiva* (1678; Fig. 47), not only did he expound the law that bears his name, namely

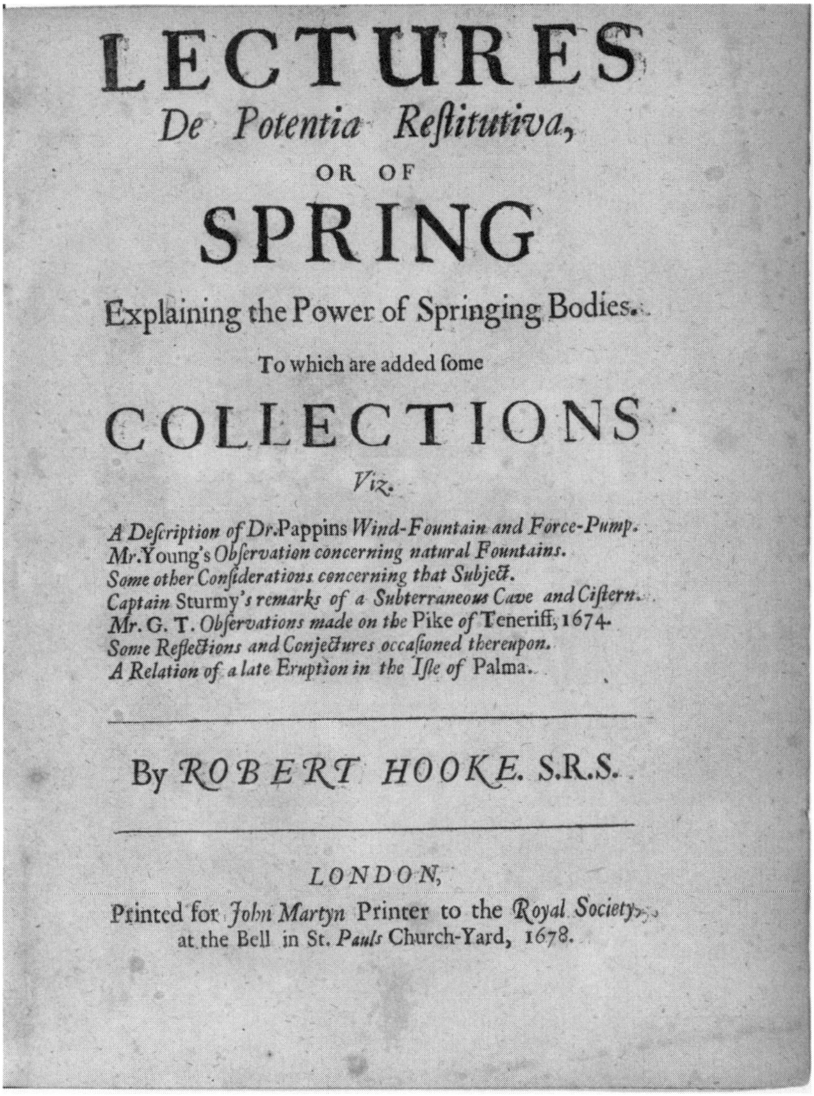

LECTURES
De Potentia Reſtitutiva,
OR OF
SPRING
Explaining the Power of Springing Bodies.
To which are added ſome
COLLECTIONS
Viz.
A Deſcription of Dr.Pappins *Wind-Fountain and Force-Pump.*
*Mr.*Young's *Obſervation concerning natural Fountains.*
Some other Conſiderations concerning that Subject.
Captain Sturmy's *remarks of a Subterraneous Cave and Ciſtern.*
Mr. G. T. *Obſervations made on the* Pike *of* Teneriff, 1674.
Some Reflections and Conjectures occaſioned thereupon.
A Relation of a late Eruption in the Iſle of Palma.

By *ROBERT HOOKE.* S.R.S.

LONDON,
Printed for *John Martyn* Printer to the *Royal Society,*
at the Bell in St. *Pauls* Church-Yard, 1678.

Fig. 47 The title-page of Hooke's *Lectures de Potentia Restitutiva* (1678). (Museum of the History of Science, Oxford.)

that 'the Power of any Spring is in the same proportion with the Tension thereof'; he also further developed the ideas about 'congruity' that had arisen from his collaborative work with Boyle and had initially been set out in his *Attempt at the Explication of the Phenomenon* of 1661 and in *Micrographia*, into what has been described as 'a unified theory of nature'.[80] Stating that 'By *Congruity* and *Incongruity* then I understand nothing else but an agreement or disagreement of Bodys as to their Magnitudes and motions,' he went on:

> Those Bodies then I suppose congruous whose particles have the same Magnitude, and the same degree of Velocity, or else an harmonical

> proportion of Magnitude, and harmonical degree of Velocity. And those I suppose incongruous which have neither the same Magnitude, not the same degree of Velocity, nor an harmonical proportion of Magnitude nor of Velocity.

He combined this with developing the ideas about the significance of vibrations, pendulums, and other periodic phenomena, using these to account for the cohesion of all sensible bodies.[81]

In his 'Lectures of Light'—given as Geometry Lectures at Gresham College in 1680–2—Hooke developed ideas initially expounded in *Micrographia* about the mechanical nature of light, seeking 'to make the manner of its Operations mechanically and sensibly intelligible'.[82] In them, as was appropriate to his Gresham audience, a survey and assessment of previous studies of the subject was interspersed by all sorts of piecemeal insights and digressions on related themes, including human perception, a topic that led him on to expound his theory of memory and the workings of the human mind.[83] His 'Discourse of the Nature of Comets' again represented lectures dating from the early 1680s; these were stimulated by the spectacular appearances of comets in the skies at that point, which led Hooke to build on *Cometa* by using his observations of the new celestial appearances as a basis for an explication of the phenomena involved. Another series of lectures, given in 1683–5, dealt with navigation and astronomy, again covering ancient and recent theories, and giving a survey of celestial and terrestrial mensuration, reverting to Hooke's favourite theme of instrumentation in giving an account of it.[84] Indeed, it is worth alluding here to another interest of Hooke's which much preoccupied him in the 1680s, namely the large-scale cartographic projects in which he was associated with the bookseller, Moses Pitt, and others, though disappointingly little came to fruition.[85]

Perhaps above all, there are Hooke's extensive geological writings, reflecting a long-standing interest on his part which, it has been suggested, may have gone back to his childhood experiences on the Isle of Wight.[86] These began with a set of lectures given in 1667–8, in which Hooke took up his speculations on such topics that had appeared in *Micrographia*. They then continued throughout the rest of his life, many of them being given as Cutlerian Lectures and hence being preserved by Hooke and then published by Richard Waller in the *Posthumous Works*. The series opens with a group of exquisite illustrations of fossils, analogous to the engravings in *Micrographia*, illustrating the intricacy and diversity of fossils by implicit analogy to the natural bodies illustrated in his earlier book (Fig. 48; see also Fig. 54).[87] Essentially, as we have already seen in connection with *Micrographia*, Hooke was reluctant to

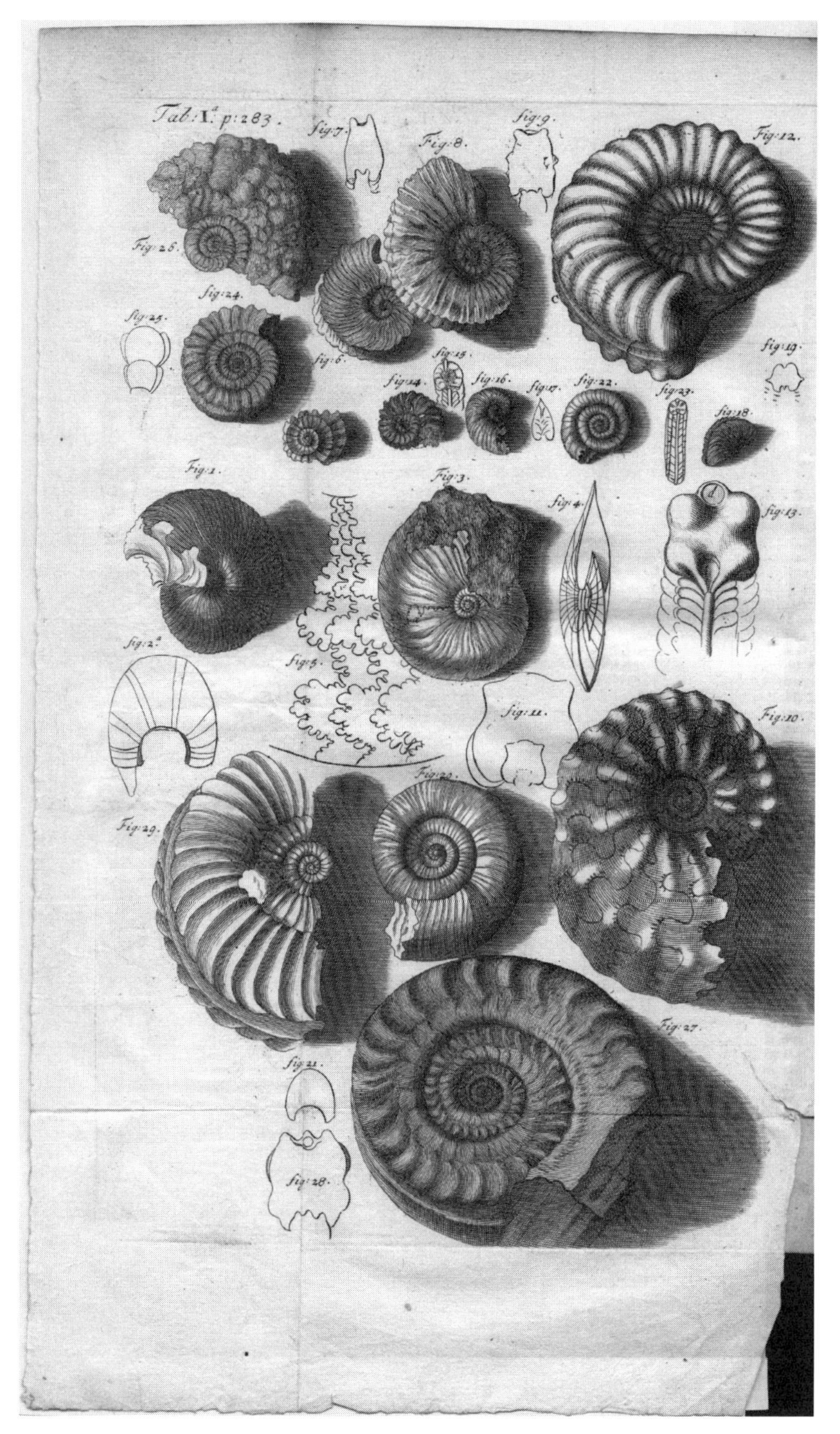

Fig. 48 One of the engravings of fossils reproduced in Hooke's *Posthumous Works*, displaying the same attention to detail shown in *Micrographia*. (Museum of the History of Science, Oxford.)

believe that fossils were created by some 'plastic virtue'. Instead, he argued that they were the remains of genuine natural bodies, often marine creatures which had been petrified and deposited deep in the earth by natural processes which it was possible to elucidate, including earthquakes.

'The concurrent Causes assisting towards the turning of these Substances into Stone,' he believed,

> seem to have been one of these, either some kind of fiery Exhalation arising from subterraneous Eruptions or Earthquakes; or secondly, a Saline Substance, whither working by Dissolution and Congelation, or Crystallization, or else by Precipitation and Coagulation; or thirdly, some glutinous or bituminous Matter, which upon growing dry and setling grows hard, and unites sandy Bodies together into a pretty hard Stone; or fourthly, a very long continuation of these Bodies under a great degree of Cold and Compression.

In other words, Hooke now proposed a mechanism for the process of petrifaction postulated in *Micrographia*. He also thought it 'not improbable, that the tops of the highest and most considerable Mountains in the World have been under Water, and that they themselves most probably seem to have been the Effects of some very great Earthquake', arguing that 'a great part of the Surface of the Earth hath been since the Creation transformed and made of another Nature; namely, many Parts which have been Sea are now Land, and divers other Parts are now Sea which were once a firm Land.' Equally boldly, he argued that 'there have been many other Species of Creatures in former Ages, of which we can find none at present; and that 'tis not unlikely also but that there may be divers new kinds now, which have not been from the beginning'—a striking and prescient notion, foreshadowing later evolutionary views, though there is no evidence that Hooke influenced these.[88]

In his various lectures, he adduced all sorts of evidence in support of these views. In particular, he sought to provide a mechanism for the eruptions that, according to his theory, had led to parts of the earth being submerged beneath the ocean while other, formerly submerged areas became dry land. Combining his geological with his cosmological views, he found this in his postulate of a shift in the earth's centre of gravity over the years. This resulted in changes in the earth's axis of rotation and of the position of the poles, producing earthquakes, inundations, and sedimentation, and this dynamic view of the earth's development, combining both geometrical and historical perspectives, represents one of Hooke's most striking and original contributions to knowledge.[89]

On the other hand, not all were convinced by Hooke's characteristic boldness in putting forward theories of this kind, particularly such Oxford scientists as Robert Plot and John Wallis, who were 'not forward, to turn the world upside down (for so 'twas phrased) to serve an hypothesis, without cogent reason for it; not only, that possibly it might be so; but that indeed it hath been so'. They also argued that such notions were not borne out by evidence from 'the History of all Ages': since all the changes that Hooke postulated were construed as fitting within the time-scale of world history provided by the Bible, it stood to reason that there should be historical records of them.[90]

As a result Hooke's ideas developed interestingly during his career, and Rhoda Rappaport has argued convincingly that this represented the result of Hooke's need to confront counter-arguments like those of Wallis and Plot. In response, Hooke looked for evidence from ancient records which might bear out his theory against its detractors, in particular seeking to prove that certain classical texts were euhemerized accounts of cataclysms like those that he thought the earth must have experienced. Thus he interpreted the Wars of the Titans as recounted by Ovid in his *Metamorphoses* as referring to an earthquake, for instance, and by this process he sought to find veiled historical testimony to the cataclysms which he considered must provide a fully mechanical explanation of the earth's development.[91]

Magician or 'Modern Physicist'?

What should be clear by now is the sheer range of Hooke's science, and the ingenuity that he brought to bear on every phenomenon that he studied. It will also be apparent that, in order to try to make sense of phenomena, he went beyond a strict interpretation of the mechanical philosophy to draw on ideas from other traditions, including those of Paracelsus, which could be justified on the grounds that they made better sense of observed phenomena than did doctrinaire mechanism. Indeed, it has been argued that Hooke went so far beyond this that he challenges a strict definition of the mechanical philosophy, and that in fact he had much in common with the tradition of 'natural magic' that had flourished in the Middle Ages and in the Elizabethan period, dealing with phenomena that existed in nature but could not be explained according to the terms of reference of scholastic science. It is certainly true that Hooke seems to have felt sympathy for such figures as Roger Bacon in the thirteenth century and John Dee in the sixteenth, both of them figures who in his view had been unfairly pilloried for their unusual but perceptive views.[92] Indeed, the natural magical tradition can be seen as one of the sources of Hooke's combination of technical virtuosity, empiricism, and eclecticism.

It is undoubtedly true that, for a mechanist, Hooke sometimes uses surprisingly non-mechanistic language—for instance, using 'sympathy' and 'antipathy' as synonyms for 'congruity' and 'incongruity' in *Micrographia*.[93] Moreover, some of his notions had definite non-mechanical overtones, perhaps particularly the significance that he attached to 'harmony' in his views on the congruity of particles, his invocation of 'active principles' both in the animate and the inanimate creation, and his description of light and gravity as the 'Souls of the greater Bodies of the World'.[94] To this extent, the view of Hooke which has been put forward by John Henry and Penelope Gouk is convincing. It certainly illustrates how we need a definition of the mechanical philosophy broad enough to encapsulate this, and suggests that an unduly strict demarcation between the mechanical and magical traditions in Hooke's period may be ill-advised.

Yet what needs to be emphasized in Hooke's case—and what is in danger of being obscured in the writings of such scholars—is that Hooke himself was consciously and explicitly anti-magical. In other words, by his time the 'magic' of the natural magical tradition had become vestigial. A telling instance of the limits to his sympathy for the activities of a man like Dee is provided by a strange Cutlerian Lecture that Hooke gave on the Elizabethan magus in 1690. The gist of his lecture was that, having decided that Dee seemed to be 'an extraordinary Man, both for learning, Ingenuity and Industry', he decided to re-examine Dee's 'Book of Spirits'—the account of his seances with angels published in 1659 by the divine, Meric Casaubon—which, if 'understood according to the plain literal Meaning', seemed to comprise 'a Rhapsody of incoherent and unintelligible Whimsies of Prayers and Praises, Invocations and Apparitions of Spirits, strange Characters, uncouth and unintelligible Names, Words and Sentences, and Relations of incredible Occurrences'.[95] Dee had seen these conversations as representing a synthesis of natural philosophy and spiritual revelation, though others (including Casaubon) considered them evidence of dabbling in a highly dangerous form of magic.[96] Hooke sidestepped the issue by arguing that the book was in fact written in code, comprising a 'concealed History' of 'Nature and Art' (hence accounting for its relevance to the Cutlerian rubric). In his view, Dee had used cryptography 'that he might the more securely escape discovery, if he should fall under suspicion as to the true Designs of his Travels'.[97] What is revealing is his presumption (in the words of the Royal Society's Journal Book) that it was unthinkable that the texts could actually record seances with angels as against examples of cryptography, for Dee was 'a very extraordinarily Knowing Man of his time, and not to be supposed capable of such incoherent ridiculous fancies as are in appearance contained in that Book'.[98] It does not seem to have occurred to him that, if Dee *was* using code, the craziest thing to have done would have been to encode his messages in a form that would have brought down the wrath of those with moralistic qualms about angel communication. Arguably, there is a slightly obtuse quality to this presumption, not wholly dissimilar to his 'interpretation' of classical myths as accounts of earthquakes, on the grounds that 'true Histories' had been converted into 'Romantick Fables'.[99]

Equally striking, both in his geology and in his attitude to magic, is his rejection of obscurity and obfuscation. Thus, in one of his later geological lectures, he claimed that the explanation of fossils as

> formed by a plastick faculty that Imitates in Sport the formes of the animall and vegetable kingdomes tells me noe more then Hesiod does in his Theogonia that every particular effect had a particular daimon or deity that took care of Producing it. whereas if I can be informed certainly that these productions are nothing els but the Reall shells, bones, teeth, or parts of Animals or Vegetables preserved by meanes of a petrifying water or Juice . . . this will give me a great stock of further inquirys and Discover the causes and Reasons of a great many other phænomena.[100]

His reaction to occultist explanations of natural phenomena was comparable, as is seen most clearly in an excursus in one of his 1670s Cutlerian Lectures, *Lampas: or, Descriptions of Some Mechanical Improvements of Lamps & Waterpoises* (1677; Fig. 49). In it, he took issue with the Cambridge Platonist, Henry More, criticizing More's invocation of a so-called 'hylarchic spirit' to explain the hydraulic phenomena of which he had given an account. Hooke argued:

> how needless it is to have recourse to an Hylarchick Spirit to perform all those things which are plainly and clearly performed by the common and known Rules of Mechanicks, which are easy to be understood and imagined, and are most obvious and clear to sense, and do not perplex our minds with unintelligible Idea's of things, which do no ways tend to knowledge and practice, but end in amazement and confusion.

He continued by drawing a clear distinction between his outlook and More's 'magic', mockingly enquiring:

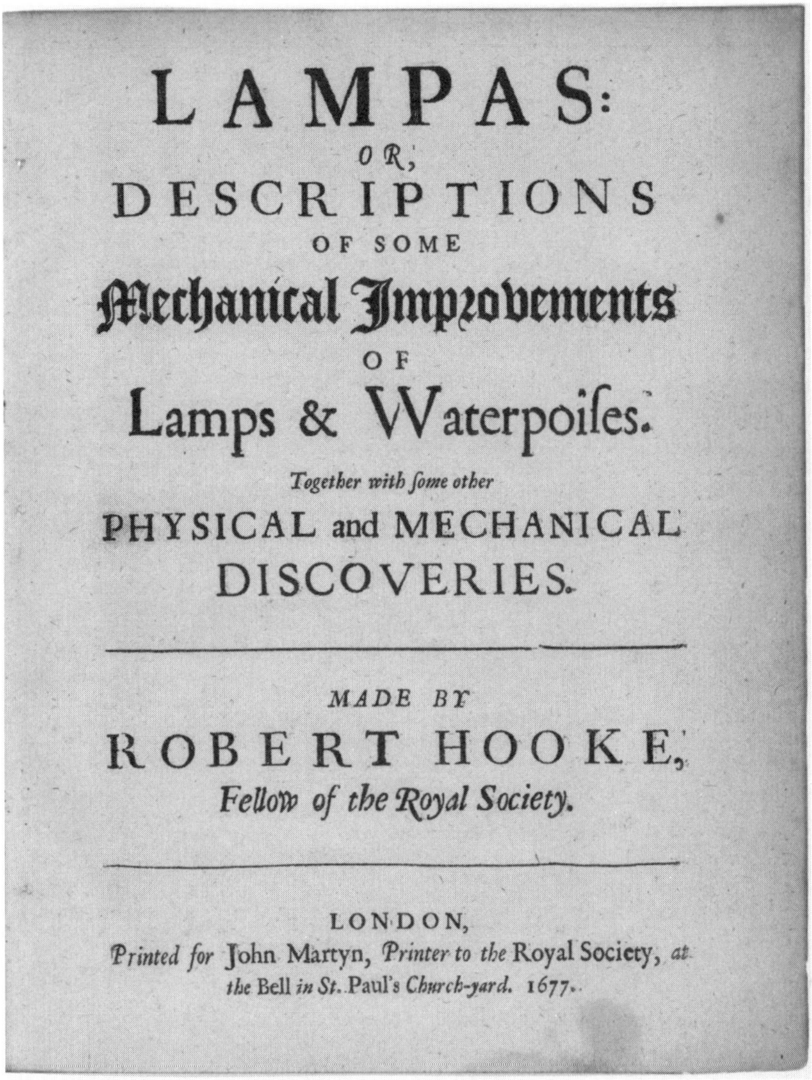

LAMPAS:
OR,
DESCRIPTIONS
OF SOME
Mechanical Improvements
OF
Lamps & Waterpoiſes.
Together with ſome other
PHYSICAL and MECHANICAL
DISCOVERIES.

MADE BY
ROBERT HOOKE,
Fellow of the Royal Society.

LONDON,
Printed for John Martyn, *Printer to the* Royal Society, *at the* Bell *in St.* Paul's *Church-yard.* 1677.

Fig. 49 The title-page of Hooke's *Lampas: or, Descriptions of Some Mechanical Improvements of Lamps & Waterpoises* (1677), which includes Hooke's attack on Henry More. (Museum of the History of Science, Oxford.)

> For supposing the Doctor had proved there were such an Hylarchick Spirit, what were we the better or the wiser unless we also know how to rule and govern this Spirit? And that we could, like Conjurers, command this Spirit, and set it at work upon whatever we had occasion for it to do. If it were a Spirit that Regulated the motion of the water in its running faster or slower, I am yet to learn by what Charm or Incantation I should be able to incite the Spirit to be less or more active, in such proportion as I had occasion for, and desired; how should I signifie to it that I had occasion for a current of water that should run eight Gallons in a minute through a hole of an Inch bore?

As he recapitulated:

> This Principle therefore at best tends to nothing but the discouraging Industry from searching into, and finding out the true causes of the Phenomena of Nature: And incourages Ignorance and Superstition by perswading nothing more can be known, and that the Spirit will do what it pleases.[101]

He echoed this in his 'Discourse of the Nature of Comets', where he contrasted the kind of explanations that he furnished with supernaturalist ones in terms of the influence of angels or devils, which he condemned as 'the Subterfuges of Ignorance, and the want of Industry'.[102] Hence, Hooke was quite self-conscious in rejecting overtly supernatural explanations, whatever his ambivalence may have been towards 'vitalistic' views of the kind with which many at the time modified their adherence to the mechanical philosophy to take into account non-mechanical powers and thus make sense of the workings of nature.[103] What needs to be emphasized is that Hooke's stress was always on clarity and intelligibility. The fact that (like other contemporaries) he was not a dogmatic or narrow mechanist did not alter his intolerance of systems of thought which he considered obfuscating, if not nonsensical—such as magic.[104]

Yet what I think Hooke did inherit from the 'natural magical' tradition was its tantalizing promise to be able to achieve the ostensibly impossible. It is almost as if Hooke had effected the transition from magician to a certain kind of modern theoretical physicist, in the way in which he played on the partial revelation of amazing truths about nature to which he claimed that he was privy. In other words, Hooke was a 'scientist' in a full, modern sense, yet this was not exclusive of his being something of a 'wonder-monger'.[105]

A case in point is provided by the possibility of human flight, which had been a traditional ambition for natural magicians.[106] Hooke took an interest in this from an early age; he claimed to have invented thirty ways of flying while still at Westminster, and he pursued this aim under the aegis of John Wilkins at Oxford: indeed, the possibility of flight was tantalizingly hinted in two places in *Micrographia*.[107] Thereafter, this trope recurs in his *Diary*, on 8 October 1674, for instance, when he 'told Sir Robert Southwell that I could fly, not how', while on 11 February the following year there was a whole discussion on the topic at a meeting of the Royal Society. This was stimulated by a paper by Dr William Croone about the muscular structure of a duck's wing, which led him to note the 'quite different structure of the body of man from that of birds, and thence concluded his utter unfitness for flying'. For this 'gave occasion to some of the members to remark, that what nature had denied to the body of man, might be supplied by his reason and by art', and Hooke

> intimated, that there was a way, which he knew, to produce strength, so as to give to one man the strength of ten or twenty men or more, and to contrive muscles for him of an equivalent strength to those of birds. He hinted likewise, that a contrivance might be made of something more proper for the feet of man to tread the air, than for his arms to beat the air.

(In the account of this in his *Diary*, Hooke extrapolated further, noting how he 'Told my way of flying by vanes tryd at Wadham,' continuing: 'Told Dr Wrens way of kites, of the unsuccessfulness of Powder for this effect, and what tryalls and contrivances I had made.' Hooke records in the *Diary* on another occasion how Wren 'liked' a related idea that he outlined to him, suggesting a comparable idea of his own.)[108] What is significant here is that it is plain from such details as Hooke gave that he was thinking of a mechanical, naturalistic method of achieving human flight. Yet the very way in which he only partially divulged what he had in mind—as when he told Southwell that he could fly but not how—preserved something of the mystique which magicians had long enjoyed, leaving the whole thing titillating and vague. It is no wonder that the dramatist, Thomas Shadwell, picked up on this in his satire of contemporary science in general and Hooke in particular in *The Virtuoso* of 1676: as its hero, Sir Nicholas Gimcrack, says

> You know a great many virtuosos that can fly, but I am so much advanc'd in the art of flying that I can already outfly that ponderous animal call'd a bustard. . . . Nay, I doubt not but in a little time to improve the art so far, 'twill be as common to buy a pair of wings to fly to the world in the moon as to buy a pair of wax boots to ride into Sussex with.[109]

'A Strange Unsociable Temper': Hooke's Intellectual Personality

At this point I want to try to assess Hooke's intellectual personality as a whole, and flying presents a good example of a further characteristic of Hooke as an intellectual, which was more of a weakness than a strength. For this is a topic on which he never wrote a treatise in which he worked out his ideas in full. Waller tells us that Hooke's interest in the subject was 'confirm'd by several Draughts and Schemes upon Paper, of the Methods that might be attempted for that purpose', which 'I have now by me, with some few Fragments relating thereto, but so imperfect, that I do not judge them fit for the Publick.'[110] This is quite plausible: after all, it was essentially a matter of technical ingenuity at which Hooke was a master. But the fact that Waller thought the scraps he found not worthy of publication indicates that nothing was ever properly worked out, and this failure to carry things through was typical of Hooke's work more generally—reminiscent, indeed, of Leonardo da Vinci himself.

Occasionally, Hooke did deal with things systematically, thereby demonstrating the exemplary method laid out in the *General Scheme* which I outlined earlier: a case in point would be his geology lectures, as studied by David Oldroyd.[111] More often, however, he did not, instead taking an almost capricious line regarding what was possible or achievable, as expressed in the preface to his published Cutlerian Lectures, where he wrote:

> as there is scarce one Subject of millions that may be pitched upon, but to write an exact and compleat History thereof, would require the whole time and attention of a mans life, and some thousands of Inventions and Observations to accomplish it: So on the other side no man is able to say that he will compleat this or that *Inquiry*, whatever it be, (The greatest part of Invention being but a luckey bitt of chance, for the most part not in our own power, and like the wind, the Spirit of Invention bloweth where and when it listeth, and we scarce know whence it came, or whether 'tis gone.) 'Twill be much better therefore to imbrace the influences of Providence, and to be diligent in the inquiry of every thing we meet with. For we shall quickly find that the number of considerable Observations and Inventions this way collected, will a hundred fold out-strip those that are found by Design. No man but hath some luckey hitts and useful thoughts on this or that Subject he is conversant about, the regarding and communicating of which, might be a means to other Persons highly to improve them.[112]

To this has to be added the extent to which Hooke's endless speculativeness made him allusive, even slightly elusive, about what he could explain, what he could prove, and what he could invent, intriguing his audience with tantalizing ideas which he never fully worked out. This is seen even in *Micrographia*, for, despite his strictures on Descartes for allowing his theories to run ahead of the data that he had to prove them, Hooke is far from free of such tendencies himself. He is constantly saying 'Which Explication I could easily prove, had I time; but this is not a fit place for it,' or how 'In this place I have onely time to hint an *Hypothesis*, which, if God permit me life and opportunity, I may elsewhere prosecute, improve and publish,' or how he was prompted 'to propose certain conjectures, as Queries, having not yet had sufficient opportunity and leisure to answer them my self from my own Experiments or Observations'.[113] This was combined with a grandiloquence about his claims which was also potentially problematic. There are thus repeated claims to originality, as where he writes how from an experiment that he had adduced, 'we may learn, that which has not, that I know of, been publish'd or hinted, nay, not so much as thought of, by any.'[114]

In many respects *Micrographia* gives a portrait of Hooke in his heyday, revelling in his sheer fecundity of invention and his ability to come up with brilliant ideas at every turn, whether or not he ever had time to work them out. Moreover, it could be argued that this was well suited to the kind of performative role that Hooke played at the Royal Society, where the premium was on titillation and the expectation of great things that might be achieved, and where, so long as the audience was satisfied, the systematic working-out of ideas could be postponed or delegated to collaborative effort of the kind advocated in his *General Scheme*. To have witnessed Hooke in action at meetings of the Royal Society must have been truly exciting, giving an exhilarating sense of being in the presence of genius ever on the verge of a breakthrough. To a significant extent, *Micrographia*, in which many of these presentations were written up, captures this. But it must have been much more generally true of Hooke's patter as he presented his work, even though the minutes as published by Birch are often rather unrevealing from this point of view. A clue to how Hooke actually presented his findings is arguably given by the contrast between the minutes' rather deadpan, matter of fact record of the demonstration of 'some experiments of descending bodies in water' on 5 and 12 October 1664 and Hooke's own report on them in one of the profuse letters to Robert Boyle that he wrote during Boyle's absence from London in 1663–5:

> amongst the rest (which are not yet brought to an exactness, and therefore I shall not till then trouble you with them) there was this very considerable discovery (for I do not find it was discovered, or so much as supposed before, but rather the clean contrary believed and builded on) that of two bodies of equal weight of the same wood, of the same shape, as to that part, which did as it were cleave the water (which was conical, being a cone, whose basis was three inches diameter, and whose altitude was two) that body did descend the fastest through the water, which had the upper end flat, and that body the slowest, which had the hindmost end sharp.[115]

The *Diary* gives an equally strong sense of discovery in the 1670s, and of course it was precisely this quality that impressed such supporters of Hooke as John Aubrey, who became an intimate friend of his in that decade. Aubrey wanted to embellish his *Brief Life* of Hooke with 'a catalogue of what he hath wrote; and as much of his inventions as I can. But they are many hundreds; he believes not fewer than a thousand. 'Tis a hard matter to get people to doe themselves right.' Aubrey also wrote, commenting on Hooke's 'prodigious inventive head': 'Now when I have sayd his inventive faculty is so great, you cannot imagine his memory to be excellent, for they are like two bucketts, as one goes up, the other goes downe.'[116]

If Hooke had been happy simply to continue his stream of ingenious speculations, often leaving them to others to carry through, all might have been well. But unfortunately he was not. For he also showed a strong possessiveness about his ideas, a characteristic he shared with other natural philosophers of the day, even if ultimately Hooke took this further than most. Indeed, there was a paradoxical contradiction between the Baconian ethos of collaboration and cooperative endeavour to which these men paid lip-service, and their actual insistence on intellectual copyright.

This was true not least of Boyle, who was extremely prone to accuse others of plagiarizing his ideas, particularly in his later years, as younger researchers entered fields which he had explored long before; in earlier days, he himself had shown little compunction about adopting others' ideas without proper acknowledgement. For instance, in 1680 he was responsible for a public attack on the French savant, Jean Baptist Duhamel, for unacknowledged use of his work in his student text, *Philosophia vetus et nova ad usum scholæ accommodata*, one chapter of which comprised 'an Abridgment of a great part of the Notions, Experiments and Ratiocinations of the *Sceptical Chymist*, without any mention there made, either of the great and famous Authors Name, or his Book in which they first appear'. The result, as Boyle's 'publisher' observed on his behalf, was that 'if the Reader were not advertis'd, he might easily suspect, that *Mr Boyle* had not lent to, but borrowed of an Author, who appears so capable of enriching the Curious with excellent things of his own.'[117]

As Hooke grew older, he similarly expected a proper deference for his ingenuity and achievement. Now here a problem was created by the habit of leaving investigations unfinished and unpublished—if not unrecorded—that had been encouraged by his role and intellectual style. Indeed, if it had not been for the contractual obligations already referred to, he would have written down even less, and it is revealing how often, in his later priority disputes, he was to appeal to the Royal

Society's records as the place where a written account of his achievements ought to be available.[118] This was truly an Achilles' heel in the dispute with Newton over the idea of universal gravitation that was central to the *Principia*. For, as Edmond Halley put it in a letter to Newton concerning Hooke's claims, 'nothing therof appearing in print, nor on the Books of the Society, you ought to be considered as the Inventor; and if in truth he knew it before you, he ought not to blame any but himself, for having taken no more care to secure a discovery, which he puts so much Value on.'[119]

Things were made worse by another of Hooke's intellectual traits, his evident presumption of his undisputed brilliance, which made him ill-equipped to handle the arrival on the scene of someone equally, if not more, brilliant, in the form of Newton. The way in which this made him intolerant of others' ideas is clear from his initial exchange with Newton over his celebrated paper on colours, published in *Philosophical Transactions* in 1672. In this case, there were various mitigating factors. Thus it did not help that (as Hooke was to complain to Lord Brouncker retrospectively) 'I had not above three or 4 hours time for the perusall of Mr Newtons paper and the writing my answer.'[120] In addition, the version of Newton's paper which he saw contained an assertion of the superiority of mathematical deduction over experiment which was bound to cause offence to him and many in the Royal Society, as Oldenburg recognized in suppressing the passage in question from the published version (it was arguably all the more sensitive for Hooke because of his own ambivalence between empiricism and hypotheticalism).[121] And linked to this was the issue of whether Newton's theory was as incontrovertibly proved by the experiments he adduced as he claimed. Nevertheless, there was a dog-in-a-mangerish aspect to Hooke's failure seriously to consider the younger scholar's epoch-making claim that white light was itself an amalgam of coloured light, and his dismissal of Newton's findings as perfectly explicable in terms of the hypothesis concerning light outlined in *Micrographia* (in which he was in any case proved wrong by Newton's devastating reply, published in *Philosophical Transactions* later that year).[122]

Equally unfortunate, and equally characteristic, considering that by this time he was trying to make things up somewhat, was his comment on Newton's second paper on colour and light in 1676, that he was 'extremely well pleased to see those notions promoted and improved which I long since began, but had not time to compleat'.[123] He thus presented Newton's experiments as a working out of his own ideas in a manner calculated to enrage the younger man, and he was to display a similar attitude later in relation to the writings on light of the Dutch natural philosopher, Christiaan Huygens, and others.[124] The exchange with Newton in the 1670s also brought out other aspects of Hooke's intellectual style which Newton was able to excoriate, including his characteristic assertion that 'I may possibly hereafter shew some proof' of his views on how telescopes might be improved, or his blindness to the fact that he too was often refining insights taken from Descartes or other thinkers; it was this, of course, on which Newton was to reflect in his famous passage about how 'If I have seen further it is by standing on the sholders of Giants.'[125]

What made matters worse was Hooke's imperfect sense of the difference between ideas that he had sketched on the back of an envelope, as it were, and ones that he had fully formulated. This problem became intense in his confrontation with Newton over the intellectual copyright of the idea of universal gravitation, but it had arguably been equally prominent in a further dispute of the 1670s that fuelled Hooke's sense of grievance over the way in which (in his view) his ideas were plagiarized by others. Here, I am referring to the famous controversy over the invention of the spring-balance watch, which Jim Bennett has already dealt with. As he has shown, in 1675 Huygens divulged his own device to Oldenburg, encouraging him to acquire the English patent for it, whereas Hooke claimed that this was his own invention of a decade earlier. Though this argument is primarily related to Hooke's activity as an inventor, it is also relevant here, partly because it fuelled Hooke's sense that others were constantly laying claim to ideas that were rightly his, and partly because of the difficulty he had in actually proving that he had formulated the invention at the point he claimed he had, since the documentation he was able to produce was feeble in the extreme.[126]

Fig. 50 Isaac Newton (1642–1727), mezzotint by John Smith after Sir Godfrey Kneller's portrait of 1702. (Private collection.)

This, therefore, brings us to the vexed issue of the rival claims of Hooke and Newton to be the true originator of the inverse square law of universal gravitation (Fig. 50). As we have seen, Hooke had canvassed related ideas ever since the publication of *Micrographia*, not least at the Royal Society in 1666 and (in print) in his *Attempt to Prove the Motion of the Earth* of 1674. Moreover, the exchange of letters with Newton which occurred in 1679–80, during Hooke's brief tenure as Secretary of the Royal Society, was undoubtedly important to the evolution of Newton's thinking, as scholars have recognized ever since W. W. Rouse Ball first rediscovered the letters in the Victorian period.[127] In his letters to Newton at this point, Hooke clearly but discursively, rather than mathematically, suggested that the elliptical orbits established by Johann Kepler at the start of the seventeenth century could be explained in terms of a single attractive force operating between the planet and the sun, which is inversely proportional to the square of the distance between them. This was completely innovatory at a time when others, including Newton, were thinking in terms of a balance of inward and outward forces keeping the planets in their orbits (as in Cartesian vortexes). It is generally accepted that it was only after this correspondence that Newton began to think in terms of actions at a distance for the first time, and most would agree with the conclusion of Alexandre Koyré, in his brilliant study of the affair, that this episode played 'an important, perhaps a decisive, role in the development of Newton's thought'.[128]

Of course, this was not something that Newton was likely to accept, particularly in relation to Hooke. The grudge against him that Newton had harboured ever since their initial confrontation in 1672 was almost certainly intensified in the course of the 1679–80 correspondence, when Hooke was able to correct an error on Newton's part. Indeed, Newton was notoriously reluctant to give credit to anyone for ideas that he considered rightfully his, offering a further extreme exemplification of the possessiveness that I have already instanced in Boyle, and Hooke was particularly unfortunate in being up against such an obsessive, scheming, and ungenerous opponent. Hence when, in 1686, while the *Principia* was being prepared for publication, Halley raised the question in a letter to Newton of whether Hooke was entitled to some acknowledgement for his role, Newton was scathing in his rejection of the idea. He replied:

> Should a man who thinks himself knowing, & loves to shew it in correcting & instructing others, come to you when you are busy, & notwithstanding your excuse, press discourses upon you & through his own mistakes correct you & multiply discourses & then make this use of it, to boast that he taught you all he spake & oblige you to acknowledge it & cry out injury & injustice if you do not, I beleive you would think him a man of a strange unsociable temper.[129]

In the same letter, he also made the point that Hooke himself owed more to the Italian natural philosopher, G. A. Borelli, than he acknowledged, while Newton went on to claim that only he had been capable of working the crucial idea out in full, criticizing Hooke for 'excusing himself from that labour by reason of his other business: whereas he should rather have excused himself by reason of his inability', a telling remark in view of what I have already said about Hooke's intellectual style.[130]

As with Hooke's exchange with Newton over his paper on light, there are various complicating factors, some going wider than the dispute itself. In so far as the matter boils down to the question of whether the crucial step was to have had the brilliant idea—where Hooke believed the credit was legitimately his—or to have worked it out, a task that he claimed merely required the attentions of a mathematical drudge, this had resonances of a long-running debate between mathematicians and natural philosophers which went back to the time of Galileo and beyond.[131] Today, when mathematicians are often seen as semi-deities, Hooke's presentation of himself as the philosopher who had the idea and Newton as the mere analyst who could confirm it mathematically may seem outrageous. But it is important to make allowance for the fact that this attitude to mathematics is itself in many ways the legacy of the success of the *Principia*, whereas the older tradition—ultimately inherited from scholasticism—held that the important task was to explain how things worked in terms of causes; natural philosophy was therefore deemed superior to mathematics, which could analyse how things happened but not say why. Hence, though it clearly suited Hooke to make the claim he did, there was a real tradition behind his position. It is also revealing that Hooke's instinct was to seek to verify causal claims experimentally, rather than mathematically. This was in keeping with the traditional view: a claim of attraction between bodies might be confirmed by a physical demonstration that there is an attraction, whereas it was far from clear how much was added to this by showing that the bodies conformed to the mathematics expected if such an attraction existed.

On the other hand, there was also the extent to which Hooke had by this time begun to get himself a bad name for accusing everyone of similar plagiarism, hence bringing us back to the tone of the passage with which I began. That this was a common view of him by the early 1680s—whether justified or not—is suggested by the report of the Dublin physician, Thomas Molyneux, to his brother William when he was in England in 1683. He said that Hooke was widely regarded as 'the most ill-natured, self-conceited man in the world, hated and despised by most of the Royal Society, pretending to have had all other inventions when once discovered by their authors to the world'.[132] Indeed, there is evidence in the Newton–Halley correspondence to substantiate this lack of sympathy for him even in the Royal Society.[133] Moreover, Hooke did not help himself by making slightly absurd claims in his later lectures to the Society about what was in dispute between him and Newton—claiming, for instance, that Newton had obtained his ideas about the oval shape of the earth from Hooke, which was almost certainly not the case.[134]

To conclude, it could be said that, until he made an enemy of Newton, Hooke was his own worst enemy: undoubtedly, there were weaknesses as well as strengths in his intellectual personality. Yet, to a far greater extent, he has been the unfortunate victim of the virtual deification of Newton since the eighteenth century, which has meant that anyone who challenged his greatness has been presumed to be in the wrong. Opinions may differ on just how much credit Hooke deserves for his contribution to the Newtonian synthesis—though even a modest contribution would be grounds for celebrating Hooke in itself. But what is tragic from Hooke's point of view is the extent to which this episode has obscured the wider achievement that I have attempted to sketch in this essay. Looking back after 300 years, surely his extraordinary fertility of invention, his brilliance of thought, and the sheer range of topics to which he applied himself, form the legacy we should be celebrating as we enter the twenty-first century.

Notes

1. The text of the lecture is taken from the MS in Hooke's hand in Royal Society Classified Papers XX, no. 77. This passage, in particular, is a patchwork of deletions and insertions; for the sake of simplicity, the former have been ignored here, pending the publication of a definitive edition of this and other unpublished lectures by Hooke; the latter are shown ‹thus›. Hooke presumably intended the English words that follow the Greek (literally, 'rather blaming than imitating') as a rendering of their meaning.
2. The reference is evidently to Mayow's *Tractatus quinque* (1674).
3. Hall 1951, 224.
4. Bennett 1980.
5. Westfall 1972. See also Westfall 1969.
6. Hunter 1989, 306, 308, and ch. 9 passim.
7. It is, of course, important not to make *too* much of print publication, since we now know that many texts were circulated in manuscript in the late seventeenth century, and this was almost certainly true of Hooke's: see Love 1998. But the paucity of his printed works nevertheless contrasts with many of his contemporaries.
8. Birch 1756–7, i, 125ff., esp. 202–4; Frank 1980, ch. 6.
9. Birch 1756–7, i, 213, 215, and 216ff. passim.
10. Birch 1756–7, ii, 90–2; Bennett 1975, 39ff.; Pumfrey 1991, 28ff.; Gal 1996.
11. Forbes 1975.
12. Hunter 1989, 292.
13. Ibid., 284–5, 292ff.
14. Ibid., 300–1. For Hooke's illustration of felt-making, see Andrade 1950, plate facing p. 470.
15. Royal Society Classified Papers XX, passim, London Guildhall MS 1757, item 10.
16. Hunter 1989, 333.
17. Pugliese 1982, 10ff.; Oldroyd 1987; Hunter 1989, 337–8.
18. For a recent survey of Bacon's ideas, see Gaukroger 2001. For a useful edition of the principal texts referred to here, see Bacon 2000 (the quotation later in this paragraph is from p. 7; that on p. 120 from p. 109).
19. Hooke 1705, 18 and 1ff. passim.
20. Ibid., 8–10.

21. Ibid., 16–17. Cf. Birch 1756–7, i, 195–7, and above, pp. 82–3.
22. Hooke 1705, 21ff.
23. See Michael Hunter, '"The Discovery of Universal Nature": Robert Boyle and the Baconian Legacy', paper delivered at the Second International Bacon Seminar, Queen Mary College, University of London, September 2001.
24. Sprat 1667, 173–9; Hunter 1989, ch. 6.
25. Royal Society Classified Papers XX, passim.
26. Hooke 1705, 19.
27. Hunter 1975, 94.
28. Anstey 2002.
29. Hooke 1705, 6–7, 65.
30. Hesse 1966; Oldroyd 1972; Pugliese 1982, ch. 2.
31. Hooke 1705, 65ff.
32. Aubrey 1898, i, 411; Hooke 1665, 93.
33. Royal Society Early Letters H.3.64.
34. Hesse 1966, 77–9, 82; Oldroyd 1972, 113, 118–19; Oldroyd 1987, 146–9.
35. Hooke 1705, 331.
36. Hooke 1665, 107 and passim. For petrifaction, see Birch 1756–7, i, 463.
37. Ibid., i, 213, 397, 442, 490.
38. Hooke 1665, b2v, 28. For an annotated edition, albeit in Spanish, see Hooke 1989.
39. Ibid., c1v–2, d2v–f2, 38–9, 149ff.
40. Ibid., 79.
41. Hunter 1989, ch. 2.
42. Hooke 1665, 1–4.
43. Ibid., 142–3.
44. Ibid., 11ff., 49ff., 67ff., 96ff., 103ff., 107ff.
45. Ibid., 217ff., 243ff.
46. Ibid., 10, 51, and passim.
47. Ibid., 35, 144 and passim.
48. Ibid., 82ff., 93; Hesse 1966, 70–1.
49. Birch 1756–7, i, 491. See also ibid., i, 463, specifically on petrifaction.
50. Hunter 1989, 208. and ch. 6 passim; Wood 1980.
51. In this chapter I have deliberately avoided the vexed issue of the extent to which Hooke's social status affected his intellectual role, since it seems to me largely beside the point. Hooke seems to have been rather unfairly picked out for attention in this respect, while the similar background of other figures (e.g., Thomas Sprat) has been ignored. My own preferred emphasis is on Hooke's 'intellectual style', which seems to me a far more complex matter than can be encapsulated by a simple contrast with the aristocratic Boyle: this will become apparent below. For the chief protagonists of the view with which I disagree, see Shapin 1989, Pumfrey 1991. See also above, pp. 6–8, and below, p. 202.
52. Hall and Hall 1962, 400ff.
53. Hooke 1665, 8.
54. Hooke 1665, 57, 62 and passim.
55. Ibid., g1.
56. Ibid., 171 and passim.
57. Ibid., 13, 47ff. and passim.
58. Ibid., sig. f2v.
59. Ibid., 46.
60. Aubrey 1898, i, 411; Davis 1989.
61. See, e.g., Hooke 1665, 75.
62. Boyle 1999–2000, i, 159.
63. Ibid., iii, xii–xiii, and 83–93.

64. Hooke 1665, 12.
65. Ibid., 15. See also Ehrlich 1995, 131ff.; Lynch 2001, ch. 3.
66. Pugliese 1982, 166–8. For background, see Shapin and Schaffer 1985, 156ff.
67. Hooke 1665, 217ff. Here I principally follow the account in Pugliese, chs. 3–4, in my view the best account currently extant. For discussion of the relative roles of Towneley, Power, Boyle, and Hooke see also Andrade 1950, 459–60; Webster 1963 and 1965; Cohen 1964; Centore 1970, 58–60 (see also 119); Agassi 1977; and Shapin 1994, ch. 7 (a revised version of Shapin 1988), various of whom have come close to claiming that Boyle's Law should really be attributed to Hooke. On the other hand, it is appropriate to add a comment on the links between Hooke and Boyle, discussion of which is often vitiated through circular reasoning based on preconceptions. What is striking is the degree of parallelism in their scientific method and practice: for instance, in their espousal of a Baconian agenda of enquiry, including the compilation of 'heads' of enquiry. As far as their pneumatic views are concerned, there may have been an element of contrast, and Hooke may always have had a greater appetite for precise quantification than Boyle. But Boyle is easily caricatured by a presumption that only Hooke was capable of quantitative approaches, and matters are further complicated by Boyle's ambivalence over the use of mathematical language (and a range of other factors adduced by Agassi). In fact, Boyle proved himself perfectly capable of such work in such later publications as *Hydrostatical Paradoxes* (1666) and *Medicina Hydrostatica* (1690)—unless a split is imposed on his intellectual personality and anything Hooke-like is presumed to be due to the influence of Hooke, in a manner reminiscent of the methods of those who attribute the authorship of Shakespeare's plays to Francis Bacon and others. See also above.
68. Hooke 1665, 37, 46, 103, and 100ff., passim.
69. For background, see Frank 1980.
70. Hooke 1665, 13–14, 103–4; Frank 1980, esp. pp. 118–19, 137–8. See also McKie 1953; Turner 1956.
71. Hooke 1665, 54–7, 62–4, 69, 72–5, 79. See also Sabra 1967, ch. 7, and below, p. 154.
72. Hooke 1665, 47ff.
73. Ibid., 245.
74. Ibid., 110–11.
75. Ibid., 87–8.
76. Hall 1966, 24–7.
77. Hooke 1665, 167, 190.
78. Hooke 1679, 27–8.
79. Ibid., 223–4. For the methodological context, see Hooke 1705, 29.
80. Ehrlich 1995, 138; Hooke 1679, 333. The law had initially been divulged in cipher in *Helioscopes* (1676): Hooke 1679, 151. See also Ehrlich 1995, 138.
81. Hooke 1679, 339; Ehrlich 1995, 138–40; Henry 1989, 160.
82. Hooke 1705, 135 and 71ff. passim.
83. On this aspect of Hooke's ideas, see Singer 1976; Oldroyd 1980; Mulligan 1992; Kassler 1995, ch. 3.
84. Hooke 1705, 149ff., 451ff.
85. Taylor 1937, 1940. For a recent overview, see Johns 1998, 446ff.
86. Drake 1996, ch. 2 and passim. For the significance of Hooke's 1666 visit to the island, see Lisa Jardine's essay below. On Hooke's geology, see also Oldroyd 1972 and 1989; Ito 1988; Rappaport 1986. For a helpful attempt at a synthesis doing justice to contemporary preoccupations, see Rappaport 1997.
87. Hooke 1705, 281ff. Waller there juxtaposes his own drawings done in the Bristol area in 1687. For correspondence relating to these investigations on his part, see Trinity College, Cambridge, MS O. 11a. 1 (25–7).
88. Hooke 1705, 290–1. See also Drake 1996, 97–8.

89. In addition to the sources cited in no. 85, see also the prescient evaluation in Patterson 1949–50, 44.
90. Turner 1974, 167, 170, reprinted in Oldroyd 1989, 210–12, 220.
91. Rappaport 1986, 131–2. See also Birkett and Oldroyd 1991.
92. For Bacon see Royal Society Classified Papers XX, no. 79. For Dee, see below.
93. Hooke 1665, 16. In addition, in ibid., p. 91, he refers to the '*plastic* virtue of Nature' in a positive way, though the latter was possibly just carelessness, since elsewhere (e.g., p. 110) Hooke is critical of this concept.
94. Henry 1989, 155–6 and passim; Gouk 1999, ch. 6. For a contrasting view see Ehrlich 1995.
95. Ibid., 203–5.
96. Harkness 1999, esp. ch. 3. Though Hooke cited Trithemius concerning the use of code, he ignored the fact that Trithemius saw this as a means to highly occultist ends: see Walker 1956, 86–90.
97. Hooke 1705, 205–6. It is perhaps interesting to note here what Waller says about the provenance of this text, which offers an interesting insight into the way in which Hooke stored his papers (ibid., 203): 'This Paper was bought by a Gentleman in the said Book [i.e., Casaubon's 1659 edition of Dee], at the Auction of *Hook*'s Library, who was pleased to send it to me.' For a discussion of the fortunes of Hooke's manuscripts and the implications of this for understanding his ideas, see Patterson 1949–50, 328ff.
98. Royal Society Copy Journal Book, 7, 289.
99. Hooke 1705, 396.
100. Royal Society Classified Papers XX, no. 78 (3v).
101. Hooke 1679, 187–8.
102. Hooke 1705, 165.
103. See Henry 1986.
104. It is appropriate to clarify here the statement in Hunter 1975, 141n., concerning Hooke and alchemy. It is true that the letters cited there from Royal Society Early Letters P 1 57–60 show that in the autumn of 1681 Pascall interspersed his letters to Lodwick and Hooke about their shared interest in a universal language with an attempt to interest them in the Elizabethan alchemist, Thomas Charnock (see also Aubrey 1898, i, 162ff.). However, although Hooke's extant response, dated 8 October 1681, shows his knowledge of the alchemical tradition, it equally clearly reveals his scepticism about its claims from a mechanistic viewpoint (as also his attempt to avoid causing offence with Pascall by pleading his 'dulnesse' and 'Ignorance in this matter'): Early Letters H 3 65, copied in Royal Society Copy Letter Book Supplement, G–H, 319–23: this is clearly to Pascall although this is not stated there.
105. For this usage see Hunter 1989, 226.
106. Henry 1989, 174.
107. Aubrey 1898, i, 410; Waller's 'Life', in Hooke 1705, iv; Hooke 1665, d1v, 198.
108. Birch 1756–7, iii, 181; Robinson and Adams 1935, 125, 146, 233.
109. Shadwell 1966, 44–5. In his reference to 'the world in the moon' Shadwell also alludes to Hooke's early mentor, Wilkins, and his book on this subject.
110. Waller's 'Life', in Hooke 1705, iv.
111. Oldroyd 1972.
112. Hooke 1679, 'To the Reader'.
113. Hooke 1665, 41, 105, 190.
114. Ibid., 103.
115. Birch 1756–7, i, 473, 474–5. Boyle 2001, ii, 343.
116. Aubrey 1898, i, 411, 415.
117. Boyle 1999–2000, ix, 29–30. See also Hunter 2000, ch. 7.
118. Hall 1951, 225. Cf. Royal Society Copy Journal Book, 9, 171, where Hooke appealed to Sprat's *History*.

119. Newton 1959–77, ii, 443.
120. Ibid., i, 198.
121. Feingold 2000, 83–4.
122. Newton 1959–77, i, 110–16; Cohen 1978, 25ff. For discussion of this complex and contentious episode, see esp. Sabra 1967, 251–64; Bechler 1974; Westfall 1980, ch. 7.
123. Newton 1959–77, i, 412; Birch 1756–7, iii, 247ff.
124. Hall 1951, 222, 224.
125. Newton 1959–77, i, 111, 416.
126. See Hall 1978; Iliffe 1992; Johns 1998, 521–31. See also above, pp. 70–1.
127. Newton 1959–77, ii, 297ff.; Koyré 1952, 221.
128. Koyré 1952, 221. Again, the secondary literature is huge. For recent assessments, see Westfall 1980, 381ff.; Nauenberg 1994; De Gant 1995, 146–55.
129. Newton 1959–77, ii, 439. Cf. ibid., ii, 431ff.
130. Newton 1959–77, ii, 438.
131. Westman 1980; Dear 2001, ch. 4.
132. Anon. 1842, 320.
133. Newton 1959–77, ii, 442–3.
134. Hall 1951, 224, 227–8.

4

Hooke the Man: His Diary and His Health

LISA JARDINE

'Many things I long to be at, but I do extremely want time.'
Hooke to Boyle, 5 September 1667.

ROBERT HOOKE, WE ARE TOLD by Richard Waller, the close friend of his later years, was 'of an active, restless, indefatigable Genius even almost to the last'.[1] This final part of our four-part study of this extraordinarily multi-faceted man concentrates on the exceptional man himself, as he is revealed in his correspondence and in his fragmentary, surviving personal diaries. I have chosen to focus my discussion on two periods particularly well served by such documentary evidence: the mid-1660s, for which an unusually rich sequence of letters between Hooke and those to whom he was professionally answerable (Boyle and Oldenburg) survives; and a continuous run of private diary entries for the 1670s.

The exchange of letters gives us a rare chance in our examination of Hooke's life to plot with some precision the crowded chart of activities and obligations in which he was engaged in his heyday. The diary entries show him coping with this kind of pressure with a medical regimen, the details of which he records precisely over extended periods of time.

My argument will be that the nature of the programme of day-to-day work Hooke set himself was a demanding one for any person to undertake, in any age. For a man working not much more than fifty years after the founding father of the modern experimental scientific tradition, Sir Francis Bacon—gentleman-amateur, conducting his scientific activities in the slow-paced *otium* of enforced retirement—the fiercely driven practical and mental life he chose to lead was unusual, if not unique. To cope with the pressures and stress, Hooke developed a rigorously compartmentalized fashion of conducting all his business in everything he did, from salaried tasks to personal friendships, recording the details necessary to keep track of each commitment succinctly in his diary. Under that same pressure (I argue) he also became a habitual, systematic consumer of a wide range of more or less toxic pharmaceutical 'remedies', which produced as many unpleasant symptoms as they cured, and which ultimately accelerated his physical decline. In his prime he was a man of boundless energy, enthusiasm, and intellectual verve, a popular, gregarious frequenter of dinner-tables and coffee houses; in later years, however, he was an irritable, emaciated, physical wreck, 'nothing but Skin and Bone, with a meagre Aspect' (to quote Waller again).

Conflict of Interests

On 3 June 1663, Robert Hooke was elected a Fellow of the Royal Society.[2] His advancement from Curator of Experiments (the job he had been appointed to at the suggestion of Sir Robert Moray in November 1662) to full membership followed the granting of the Great Seal (royal approval) for the revised Charter of the Royal Society that April, and the re-election of the foundation Fellows a month later, to regularize the Society under the terms of its new Charter.[3]

Full membership of the Royal Society transformed Hooke from a simple 'operator', orchestrating equipment and demonstrations for the Society's weekly meetings, to decision-making participation in its scientific and social organization. It also complicated his relationship with his other regular, day-to-day employment, creating what amounted to a potentially serious conflict of interests. Since the mid-1650s Hooke had occupied the paid post of laboratory assistant and amanuensis to Robert Boyle, youngest son of the Earl of Cork.[4] Before he could be offered the post of Curator of Experiments, Boyle had to be respectfully asked to release Hooke from some of the duties he carried out for him. Nor was this just a matter of juggling employment priorities. There were also emotional pulls: Hooke had served Boyle since his teens (he was now 28); he had worked closely alongside him, and learned most of what he knew about chemistry and 'physic' (medical dosing) from him. (We will return to the medical knowledge shortly.)[5] Hooke remained in Boyle's salaried employ (and continued to have his residence in Boyle's household) until 1665, when his appointment to the Gresham Professorship of Geometry entitled him to accommodation within the College.[6]

The problem was that Boyle expected Hooke to be in attendance whether he was residing in London, Dorset (his country residence), or elsewhere, while the Royal Society expected Hooke's regular attendance at its weekly London meetings.

Two days after his election, on 5 June 1663, Hooke wrote to Boyle, who was visiting his sister Mary, Countess of Warwick, and her family, at Leese Priory in Essex. With elaborate and careful deference Hooke asked his aristocratic employer to agree to his putting off for a few days longer his previously arranged trip to Leese to join him:

> Ever honoured Sir,
> I have put up and sent the things you gave order for . . . and should have come away my self, but that having received a particular favour from the [Royal] Society, and also an extraordinary injunction to see the condensing engine [Boyle's air-pump] in a little order against the [Society meeting] next Wednesday, I did hope you would be pleased to dispense with my absence from attending on you for two or three days longer, till the next Wednesday be past. . . . For I remember you were pleased to say, that you thought it would be a week before the ceremony of visits would suffer you to settle about any business, and so should have little use of me till then; and if your occasions would permit a dispensation for my stay here any longer time, I should endeavour to improve the time the best way I am able to serve you.
>
> But, Sir, I make it no further my desire, than the convenience of your affairs permit, having wholly resigned myself to your disposal. Nor should I have presumed to have trespassed your commands thus far, had I not thought, that the Society might have taken it a little amiss, if, at the very next meeting, after so great an honour done me, I should be absent.[7]

To go some way towards placating Boyle, Hooke proceeded to describe the demonstration at the Society meeting for which his attendance as Curator and Operator was so essential—it was the operation of Boyle's air-pump (Fig. 51), the much-celebrated item of equipment, purpose-built for investigating respiration and the properties of air, designed to his specifications by Hooke, and which Hooke alone could reliably operate:

> There was nothing of experiment, but only a trial of the condensing engine, which only held enough to shew us, that it would not hold long enough with that kind of cement we used; for after the air was condensed into about half the dimensions, it forced its way through the cement of the covers, though laid very thick in the joints. But I think that inconvenience will be easily remedied against the next day.[8]

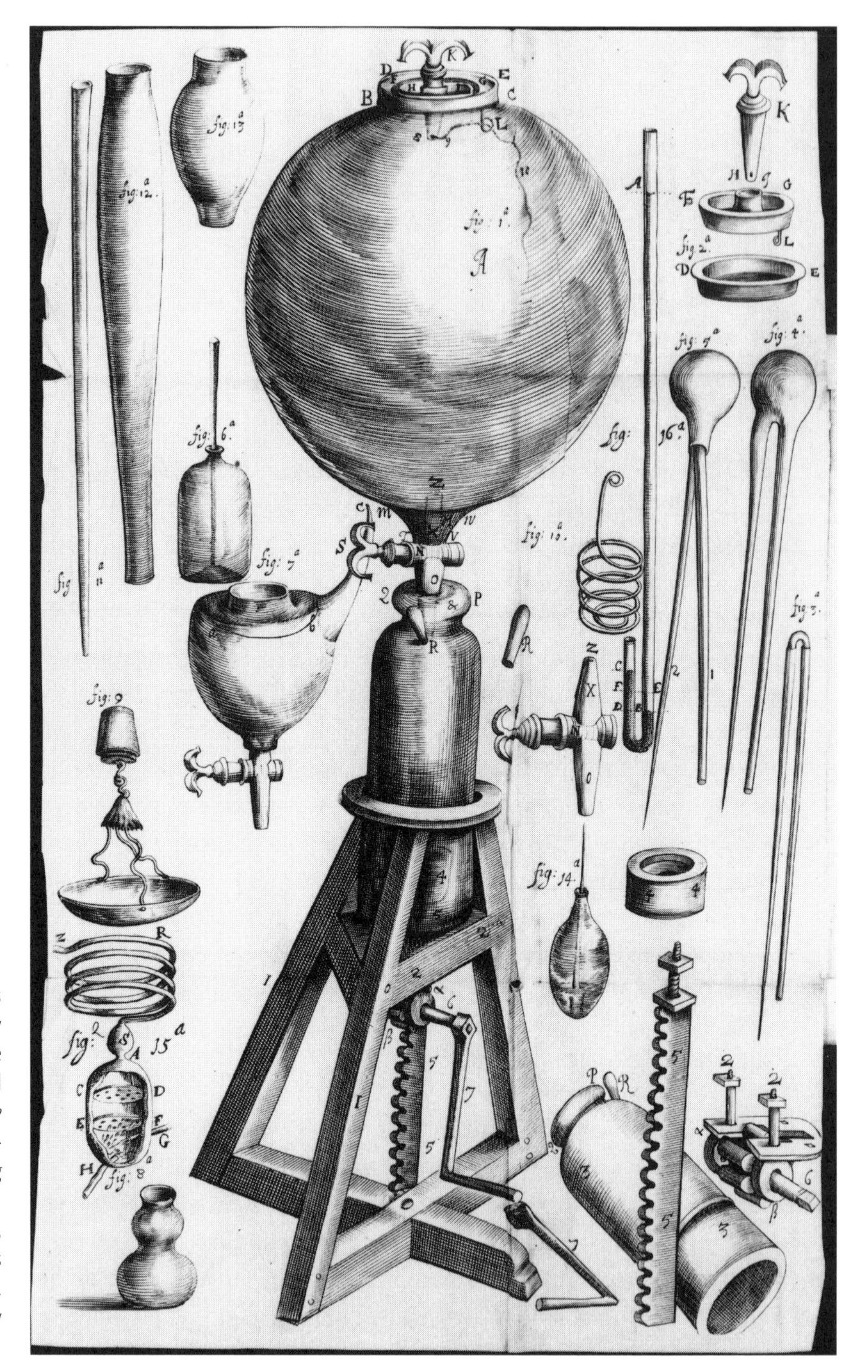

Fig. 51 Robert Boyle's air-pump, designed by Hooke and used for the experiments published in Boyle's *New Experiments Physico-Mechanicall, Touching the Spring of the Air, and its Effects* (1660), to which this plate is the frontispiece. (Museum of the History of Science, Oxford.)

Five days later, after the Wednesday meeting, Henry Oldenburg, Secretary of the Royal Society, and another long-standing member of the Boyle family's entourage (he had originally been hired as tutor to Boyle's nephew, Richard Jones), wrote to Boyle, describing with enthusiasm the events that had taken place, for which Hooke had so earnestly requested permission to remain in London:

> This afternoon we had no ordinary meeting: There were no lesse than 4 strangers [foreign visitors], ‹two› French, and two Dutch Gentlemen; the French were, Monsieur de Sorbière, and Mr Monconis; the Dutch, both the Zulichems, Father and Son [Constantijn senior and Christiaan Huygens]: all foure, inquisitive after you. They were entertained ‹first› with some Experiments, of which the bearer hereoff will give you a good account off. [9]

That 'bearer hereoff' of Oldenburg's letter—perfectly placed to give an account of the experiments conducted before the distinguished visitors—was, however, none other than Hooke himself, who could postpone his trip no longer, and had now been obliged to give way to Boyle's summons to join him in Essex (Fig. 52).

In a postscript, Oldenburg respectfully expressed the hope that Boyle would release Hooke again to recommence his Royal Society curatorial duties as soon as possible:

Fig. 52 William Faithorne's engraved portrait of Boyle with the air-pump in the background. Hooke was instrumental in the production of this image in 1664. (Ashmolean Museum.)

> Seeing the abovementioned Strangers are like to continue here yet a while, at the least some of them, the Society shall much stand in need of a Curator of Experiments; which, I hope, Sir, will the sooner procure from your obligingnes a dispensing with Mr Hook for such a publick use.[10]

It was not simply that Hooke was required to operate the air-pump effectively in the presence of distinguished foreigners—he alone possessed the necessary experimental brio and showmanship to captivate and delight dilettante visitors, combining technical virtuosity with a keen sense of theatre. On 7 May 1662, John Evelyn recorded in his diary that he accompanied the King's cousin, Prince Rupert, to a Royal Society meeting:

> I waited on Prince Rupert to our Assembly, where were tried several experiments of Mr. Boyle's Vaccuum: a man thrusting in his arme, upon exhaustion of the ayre, had his flesh immediatly swelled, so as the bloud was neere breaking the vaines, & unsufferable: he drawing it out, we found it all speckled.[11]

The man with his arm in the air-pump, prepared, as an act of bravura, to do himself physical damage in order to entertain his audience, was almost certainly Hooke.[12]

Hooke did not, in fact, return to London and to his regular duties at the Royal Society for several weeks.[13] On 22 June 1663, Oldenburg described to Boyle the formal admission of Huygens and Sorbière to the Society, but added that the air-pump experiment had failed dismally in Hooke's absence.[14] It was the beginning of July before Hooke was back in London. He and Oldenburg sent letters to Boyle by the same carrier on 3 July, both of them somewhat nervously enquiring when he would himself be returning to London, since plans were under way for a visit to the Society by the King. (Like an earlier royal visit proposed in 1661, this never in fact took place. An 'Extraordinary day' to plan this meeting was set for 6 July—in spite of Oldenburg and Hooke's urgings, Boyle missed it.)[15]

We might note that Boyle and Oldenburg were not the only ones making competing claims on Hooke's time on account of his exceptional practical scientific skills. When Sir Robert Moray proposed a project for fixing the positions of the stars in the Zodiac, in April 1663, Christopher Wren volunteered himself and Hooke (a close personal friend) to survey the constellation Taurus (Hooke included the resulting map of the Pleiades in his *Micrographia* (1665)).[16] The two men collaborated again for the Royal Society with joint observations of a comet which became visible in 1664.[17] Wren was responsible for observations in Oxford, where he held the Savilian Chair of Astronomy, while Hooke recorded and tabulated observations from London. In this case it was Wren who shuttled between the two locations. Hooke, meanwhile, tabulated the data, collated their results, wrote the subsequent Royal Society reports, and sent meticulous accounts of proceedings by letter, just as he did in relation to his activities carried out on behalf of Boyle, who was also in Oxford, largely domiciled in accommodation leased 'at Mr Crosse's house', during the same period.[18]

Feeling the Pressure

The severe outbreak of plague in London in the summer of 1665 laid a further strain upon Hooke's working arrangements, research commitments, and obligations to clients and friends. At the end of July 1665, Wren left for Paris as a member of the party accompanying Queen Henrietta Maria out of the country (partly to avoid the plague, partly on the instruction of the King, to look at the programme of building Louis XIV had recently undertaken).[19] Boyle decamped completely to his Oxford lodgings, while Sir Robert Moray and others in the court circle also moved to Oxford, to which town the King had transferred his court and administration, and where they remained until February 1666. The three leading experimentalists and operators of the Royal Society, Hooke, Sir William Petty, and Dr John Wilkins, however, under instructions from the President and Council, took a quantity of the Society's experimental equipment and materials to the Durdans estate, outside Epsom—the residence (with extensive grounds) of George, Lord Berkeley, loaned to the Society while its owner himself was with Wren's party in Paris.[20]

For some time the Royal Society had been interested in developing a fast, lightweight carriage which would give both driver and passenger a comfortable ride. In April 1665 Hooke was instructed to pursue a design proposed by Thomas Blount:

> Mr Hooke was ordered to prosecute the model of his chariot with four springs and four wheels, tending to the ease of the rider.
>
> It was likewise ordered, that the President, Sir Robert Moray, Sir William Petty, Dr. Wilkins, Col. Blount, and Mr. Hooke should be desired to suggest experiments for improving chariots and to bring them in to the mechanical committee, which was to meet on the Friday sevennight following, April 21, at the president's house.[21]

On 26 April, the same committee was instructed to assemble at Colonel Blount's house the following Monday to see his own model chariot, and to 'give an account of what they had done there at the next meeting of the Society'.[22] When the Society called its summer recess, and members dispersed to their various country locations, the chariot-testing group was exhorted to continue its experiments under ideal conditions at Durdans (where carriage racing was more practical than in the streets of London):

> The members of the society were then exhorted by the president to bear in mind the several tasks laid upon them, that they might give a good account of them at their return; and Mr. Hooke was ordered to prosecute his chariot-wheels, watches, and glasses, during the recess.[23]

When the plague outbreak was over and the Society eventually reassembled and resumed its meetings, in March 1666, it was the chariot-improvers who gave the first full report of the successful modifications carried out in Epsom:

> The president inquiring into the employments, in which the members of the society had engaged during their long recess, several of those, who were present, gave some account thereof: viz.

Dr. Wilkins and Mr. Hooke of the business of the chariots, viz. that after great variety of trials they conceived, that they had brought it to a good issue, the defects found, since the chariot came to London, being thought easy to remedy. It was one horse to draw two persons with great ease to the riders, both him that sits in the chariot, and him who sits over the horse upon a springy saddle; that in plain ground [a] 50-pound weight, descending from a pully, would draw this chariot with two persons. Whence Mr. Hooke inferred, that it was more easy for a horse to travel with such a draught, than to carry a single person: That Dr. Wilkins had travelled in it, and believed, that it would make a very convenient post-chariot.[24]

Wren—ostensibly in Paris to collect information on Louis XIV's ambitious building projects, but also taking advantage of the expenses-paid trip to conduct productive dialogues with a number of Parisian scientists—brought back comparative information on lightweight carriage construction there; he and Hooke were instructed by the Society to iron out the last defects in the new carriage's performance and report back to the Society:

Dr. Wren and Mr. Hooke being asked, what they had done in the business of chariots, since the perfecting thereof was committed to them, Dr. Wren answered, that he had given Mr. Hooke the descriptions of those, which they had in France.[25]

Novel and efficient designs for new, lightweight carriages were high on the Society's agenda, because a charge regularly levelled against the Royal Society from its inception was that its investigations yielded nothing of practical, commercial use to a nation on the lookout for wealth-creating investment opportunities.[26] The jibe that the Society's members wasted their time 'on the weighing of air' (that is, on nothing) was widely current.

The set of experiments Wilkins, Petty, and Hooke undertook at Durdans with most commercial potential was one designed to continue testing on land, with the help of those rapidly moving coaches, the viability of new marine timekeepers, based on the pendulum clock invented by the Dutch virtuoso, Christiaan Huygens (for which the Royal Society had recently been granted a lucrative patent): 'Some further experiments should be made with them, by contriving up and down motions, and lateral ones, to see, what alterations they would cause in them.' Hooke and Wilkins were also instructed to work on 'watches with springs'—an alternative form of precise timekeeper for which Hooke was currently secretly discussing his own patent with Moray, Brouncker, and Wilkins.[27]

From the Royal Society's point of view, significant sums of money depended on perfecting a marine timekeeper with either a pendulum or a balance-spring isochronous timekeeping mechanism, and these experiments were of the utmost urgency. From Hooke's own personal point of view, French claims already to have developed a balance-spring timekeeper, and the fact that he had recently been obliged to break off negotiations concerning an English patent, made it yet more crucial that he continue with this work. However, it appears that the pressure of other demands on Hooke's time at Epsom precluded his carrying out such a programme of experiments. In spite of the fact that ten years later Hooke characterized his summer 1665 work on his watch as occupying much of his time, practically and emotionally, there is little evidence of balance-spring watch-related research being carried out at Epsom.[28]

The probable reason why Hooke failed to advance the perfection of his marine timekeeper (a project dear to his own heart) was that, on top of the projects he had been explicitly instructed by Brouncker and the Council of the Royal Society to continue pursuing during the enforced recess, he was already seriously over-committed to extended programmes of experimentation in his capacity as an operator employed by each one of his now scattered clients and patrons. He arrived at Durdans having already promised to work on hydrostatics for Boyle (as Boyle's personal operator), to carry out a raft of other, specified Royal Society experiments (including air-pump experiments) with Wilkins and Petty (as leading figures on the Council of the Royal Society), and to continue telescopic observations for his astronomical work with Wren, during the latter's absence in France on royal architectural business. In 'recess' in the country, he was now obliged to juggle competing claims upon him more dramatically than ever. In the event, Boyle made the first claim.

Hooke had notified Boyle (in Oxford) in early July that he was likely to have to leave Gresham College to avoid the plague, laying careful emphasis upon the fact that the move out of London would not impede, but rather facilitate, the experimental work to which Hooke was committed on Boyle's behalf:

> I doubt not, but that you have long before this heard of the adjourning of the Royal Society, and of the increase of the sickness, which rages much about that end of the town you left. . . . We have made very few experiments since you were pleased to be present, but I hope, as soon as we can get all our implements to Nonsuch,[29] whither Dr. Wilkins, Sir W. Petty, and I, are to remove next week, I shall be able to give

> you an account of some considerable ones, we have designed to prosecute the business of motion through all kinds of mediums, of which kind Sir W. has made already many very good observations. We shall also take the operator along with us, so that I hope, we shall be able to prosecute experiments there as well almost as at London; and if there be any thing, that you shall desire to be tried concerning the resistance of fluid mediums, or any kind of experiment about weight or vegetation, or fire, or any other experiments, that we can meet with conveniencies for trial of them there; if you would be pleased to send a catalogue of them, I shall endeavour to see them very punctually done, and to give you a faithful account of them.[30]

Evidently he was convincing, and his enthusiasm caught Boyle's imagination. Boyle decided to leave Oxford forthwith, to take advantage of the circumstance which Hooke described so plausibly as especially conducive to his (that is, Boyle's) experimental requirements, in Epsom. In fact, Boyle apparently arrived first, settling himself at Durdans by 8 July.[31] In a letter of that date, he described to Oldenburg the experimental programme which he was now conducting. The fact that Boyle had decamped from Oxford suggests that this programme could best be accomplished with Hooke's assistance (with Hooke operating, and Boyle, as usual, overseeing in his capacity as patron and virtuoso). Presented with the convenience of his laboratory operator (and a junior) away from home, Boyle was happy to drop experiments involving equipment which could not be readily acquired outside London, and to get on with those which would complete his book of *Hydrostatical Paradoxes*:

> I am here about some Hydrostaticall Exercises, & if God vouchsafe me health & opportunity of staying in this place t'is like I shall ‹publish› my Hydrostaticall parad[oxes] without staying for the Appendix to the pneumaticall Booke, which for want of Receivers & other fit glasses; I have here noe conveniency to compleate.[32]

We learn more about these experiments in a letter Boyle wrote to Oldenburg in September. He was trying to write up the experiments for publication, but was being prevented from doing so by the 'crowd & hurry' of Oxford visitors:

> The ground of it is That I had by diligent Experiments (which were difficult & troublesome enough) found out the true weight of a Cubick Inch of water, that is, as much of that Liquor as would exactly fill a hollow vessell of every way an Inch in its Cavity. For this weight being

> once obtaind t'is easy to discover by the weights to be added to that scale at which the Body to be measurd hangs immersd in water, how many times it looses by that immersion as much weight as amounts to a Cubick Inch of water: soe that although by a peculiar Statera the thing may be better performd, without Calculation; yet I soe contrive the matter that only useing peculiar weights (which may be easily made & of what heavy body you please) any ‹good paire of› ordinary scales may be made with tollerable accuracy to define the magnitude of the immersd Body.[33]

In the version of *Hydrostatical Paradoxes* published by Boyle in 1666, he nevertheless returned to the difficulty of carrying out his experiments at Epsom without access to purpose-made glassware. This perhaps explains why, after a matter of weeks, and one imagines to Hooke's relief, Boyle returned to Oxford, arriving there around 6 August.

Boyle's departure presumably eased the immediate problems that Hooke may have been having as he tried to fulfil his experimental obligations for the Royal Society (Wilkins and Petty) and for his old employer (bent on completing the laboratory experiments for his forthcoming publication on hydrostatics).[34] He returned to some of the projects planned before he left London. A contemporary witness paints a picture of a particularly relaxed group of virtuosi, enjoying their recently regained freedom to control their own researches, a few days after Boyle's departure. On 7 August 1665 John Evelyn, returning to London from his family estate at Wotton, in Surrey, visited the Royal Society evacuees at Durdans and reported that the team was hard at work on those chariot experiments they had been directed to carry out:

> I returned home, falling at Woo‹d›cot, & Durdens by the way: where I ‹found› Dr. Wilkins, Sir William Pettit [Petty], & Mr. Hooke contriving Charriots, new rigges for ships, a Wheele for one to run races in, & other mechanical inventions, & perhaps three such persons together were not to be found else where in Europ, for parts and ingenuity.[35]

Boyle's return to Oxford (which was followed shortly afterwards by Petty's departure for Salisbury)[36] also meant that Hooke could turn some of his attention to ongoing work with astronomical instruments, to which he was committed as part of his observational team-work with Wren. For over a year, Wren and Hooke had been making and collating telescopic observations of the movement of the two comets which had become visible in 1664 and 1665. In this case too, a set of investigations which Hooke was instructed to undertake by the Royal Society involved him in intensive collaborative activities which had to be dovetailed with other commitments, involving other, equally demanding and socially prominent colleagues (in this case, nightly observations beginning shortly before his election to a fellowship, and continuing until Wren left for Paris). Hooke also had the job of collating the observations of other associates of the Society, submitted via individual members:

> [12 April 1665] Mr. Howard produced an account of the new comet, sent to him by his brother from Vienna; which was delivered to Mr. Hooke, to compare it with other observations.
>
> [19 April 1665]. Dr. Croune presented from Sir Andrew King a paper with a scheme of the first comet, drawn by a Spanish Jesuit at Madrid; which was delivered to Mr. Hooke to compare it with the other observations; who was also appointed to take a copy of Dr. Wren's scheme of this comet, and to return the original to the Doctor for further consideration.
>
> [26 April 1665] Mr. Howard produced some observations on the second comet, as they were sent to him by his brother from Vienna; which were recommended to the perusal of Mr. Hooke.
>
> [17 May 1665] Three accounts were brought in of the late comets; one by Dr. Wilkins concerning the first [in 1664], sent out of New England; the other two by Mr. Aerskine, concerning the latter, written from Prague and Leige: All of which were ordered to be delivered to Dr. Wren and Mr. Hooke.[37]

In the course of their observations of the comet which appeared in spring 1665, Hooke and Wren had together been developing and testing a double telescope of Wren's.[38] At Epsom, Hooke now tested and adjusted a new quadrant he had designed to be used alongside Wren's telescope. The two men's explicit hope was that, in combination, the instruments could be used to make the precise measurements necessary to determine longitude at sea by astronomical methods.[39]

On 8 July 1665 (before travelling to Epsom), Hooke had told Boyle that he was planning to experiment there with his new quadrant:

> Mr. Thompson also has sent home the instrument for taking angles, and demands two and thirty shillings. It is not quite finished, but I intend to take it with me to Nonsuch, and there to make trial of it, and adjusten it. I shewed it the last meeting of the Society, which it was very much approved of; and I hope it will be the most exact instrument, that has been yet made.[40]

On 15 August, Hooke reported to Boyle:

> One of our quadrants does to admiration for taking angles, so that thereby we are able from hence to tell the true distance between [St] Paul's and any other church or steeple in the city, that is here visible, within the quantity of twelve foot, which is more than is possible to be done by the most accurate instrument or the most exact way of measuring distances.[41]

When he and Wren were both back in London, in late February or early March 1666, Hooke and his old friend continued to work together on perfecting their longitude method (which both probably hoped would make their fortunes), adding the final touches to their procedure by the end of the summer. The two of them were scheduled to present their lunar longitude method to the Royal Society at the meeting on 12 September 1666—a meeting whose agenda was altered following the Great Fire in the first week of September.[42]

Finally, during this period outside London, Hooke took advantage of a local, man-made feature of the Epsom landscape unavailable to him in the London area—deep, abandoned wells.[43] In August he wrote to Boyle:

> I have made trial since I came hither, by weighing in the manner, as Dr. Power pretends to have done, a brass weight both at the top, and let down to the bottom of a well about eighty foot deep, but contrary to what the doctor affirms, I find not the least part of a grain difference in weight of half a pound between the top and bottom. And I desire to try that and several other experiments in a well of threescore fathom deep, without any water in it, which is very hard by us.[44]

On 26 September Hooke reported that he had indeed tried his experiments in the deeper well, though it had turned out to be 'three hundred and fifteen foot in its perpendicular depth', which was less than he had been promised, 'so that it seems no less than a hundred foot is filled with rubbish [rubble], at least it is stopped by some cross timber, which I rather suspect, because I found the weights to be stayed by them if I suffered them to descend below that depth.' He had, he reported, already conducted several experiments, including one 'of gravity', 'which upon accurate trial I found to succeed altogether as the former', and a less successful one involving lowering lighted candles, to see at what depth they were extinguished (the candles came loose from their sockets and were lost).[45] With characteristic energy, Hooke itemized a list of further well experiments he intended to carry out, so as not to waste 'such an opportunity, as is scarce to be met with in any other place I know':

> I have in my catalogue already thought on divers experiments of heat and cold, of gravity and levity, of condensation and rarefaction of pressure, of pendulous motions and motions of descent; of sound, of respiration, of fire, and burning, of the rising of smoke, of the nature and constitution of the damp, both as to heat and cold, driness and moisture, density and rarity, and the like. And I doubt not but some few trials will suggest multitude of others, which I have not yet thought of; especially if we can by any means make it safe for a man to be let down to the bottom.[46]

We may be sure that the man who would have been let down to the bottom of the well—if it had proved humanly possible—would have been Hooke himself.

A fuller record of the well-based experiments is also found in Hooke's reporting back to his Royal Society employers after the protracted recess, in March 1666. It shows that access to deep wells allowed Hooke, in his capacity as a virtuoso designer and maker of precision instruments, to develop them, as well as to conduct experimental programmes which required to be carried out under conditions as nearly identical as possible at different points above and below the surface of the earth:

> He presented a paper, which was read, containing some experiments of gravity made in a deep well near Banstead Downs in Surry; to which was annexed the scheme of an instrument for finding the difference of the weight, if any, between a body placed on the surface of the earth, or at a considerable distance from it, either upwards or downwards.[47]

The paper itself (duly registered and deposited with the Society) offers us further, precisely vivid detail concerning Hooke's Surrey experimenting with wells:

> [I]f all the parts of the terrestrial globe be magnetical, then a body at a considerable depth, below the surface of the earth, should lose somewhat of its gravitation, or endeavour downwards, by the attraction of the parts of the earth placed above it. . . .
>
> For the trial of which I had a great desire, and happily meeting with some considerably deep wells, near Banstead Downs, in Surrey, I endeavoured to make them with as much exactness and circumspection as I was able. My first trials were in a well about 15 fathoms deep,

> or 90 foot; the packthread I made use of was about 80 foot long; the bodies I weighed, or let down by it, were brass, wood, and flints; each of which, at several times, I counterpoised exactly, and hung the scales, which were very good ones, over the midst of the well, so as that the packthread might hang down to the bottom without touching the sides. The effects were these, that each of those bodies seemed to keep exactly the same gravity at the bottom of the well, that they had at the top.[48]

Here too Hooke tells the reader how he repeated his experiments 'with the like circumspection in a well of near sixty fathoms deep, where the weight, though suspended at the end of a string of about 330 feet long, seemed to continue of the same weight, that it had above, both before it was let down, and after it was pulled up'.[49] His conclusion is that instruments capable of detecting far smaller increments of change would be needed to test the alteration of gravitational pull with height, and he proceeds to describe such an instrument.[50]

Never at Rest

In the midst of Hooke's juggling commitments and experimental priorities in his out-of-town exile came a further unexpected disruption. His planned series of experiments involving deep wells was interrupted by a journey he was obliged to take away from Epsom himself, on pressing family business. Hooke's mother had died in June 1665, but the virulence of the plague meant that it had been not possible for him to travel to the Isle of Wight immediately. (In summer 1665 the local authorities in Hampshire banned all visitors from the London area, in an attempt to prevent the spread of infection.) He eventually went home in early October to busy himself with yet another set of pressing obligations. There were family properties to be sold or leased to suitable tenants, and money arrangements to be made with his sisters and his brother John (who was notoriously bad with money himself, and perhaps not to be trusted to act on Hooke's behalf).[51]

Characteristically, Hooke did not use even this enforced break from his professional employment to relax; instead, he turned it into yet another scientific opportunity for hands-on experimental research. The fossil-rich chalk cliffs around Freshwater Bay, on the south-west of the island, where he had spent his boyhood, were the perfect place to pursue his interest in geology—in fossils and theories of rock formation. In 1667, in the earliest of his 'Discourses on Earthquakes', Hooke notified his readers that it was during his trip to the Isle of Wight the previous year that he had assembled the substantial collection of fossils on which his lectures (and the exquisite drawings from which Richard Waller, editor of Hooke's *Posthumous Works*, had the engraved plates made for the published texts) were based (Fig. 53):

> I had this last Summer[52] an Opportunity to observe upon the South-part of *England*, in a Clift whose Bottom the Sea wash'd, that at a good heighth in the Clift above the Surface of the Water, there was a Layer, as I may call it, or Vein of Shells, which was extended in length for some Miles: Out of which Layer I digg'd out, and examin'd many hundreds, and found them to be perfect Shells of Cockles, Periwinkles, Muscles, and divers other sorts of small Shell-Fishes; some of which were fill'd with the Sand with which they were mix'd; others remain'd empty, and perfectly intire.[53]

Hooke proposed that rock formations had once been matter in solution which 'in tract of time settled and congealed into . . . hard, fixt, solid and permanent Forms', no longer soluble in water. As evidence for this, he cited his own observations, 'which I have often taken notice of, and lately examined very diligently':

> I made [this Observation] upon the Western Shore of the Isle of Wight. I observed a Cliff of a pretty height, which by the constant washing of the Water at the bottom of it, is continually, especially after Frosts and great rains, foundering and tumbling down into the Sea underneath it. Along the Shore underneath this Cliff, are a great number of Rocks and large Stones confusedly placed, some covered, others quite out of the Water; all which Rocks I found to be compounded of Sand and Clay, and Shells, and such kind of Stones, as the Shore was covered with. Examining the Hardness of some that lay as far into the Water as the Low-Water-mark, I found them to be altogether as hard, if not much harder than *Portland* or *Purbeck*-stone . . . Others of them I found so very soft, that I could easily with my Foot crush them, and make Impressions into them, and could thrust a Walking-stick I had in my Hand a great depth into them. . . . All these were perfectly of the same Substance with the Cliff, from whence they had manifestly tumbled, and consisted of Layers of Shells, Sand, Clay, Gravel, Earth, &c. and from all the Circumstances I could examine, I do judge them to have been the Parts of the Neighbouring Cliff foundered down, and rowl'd and wash'd by degrees into the Sea; and, by the petrifying Power of the Salt Water, converted into perfect hard compacted Stones.[54]

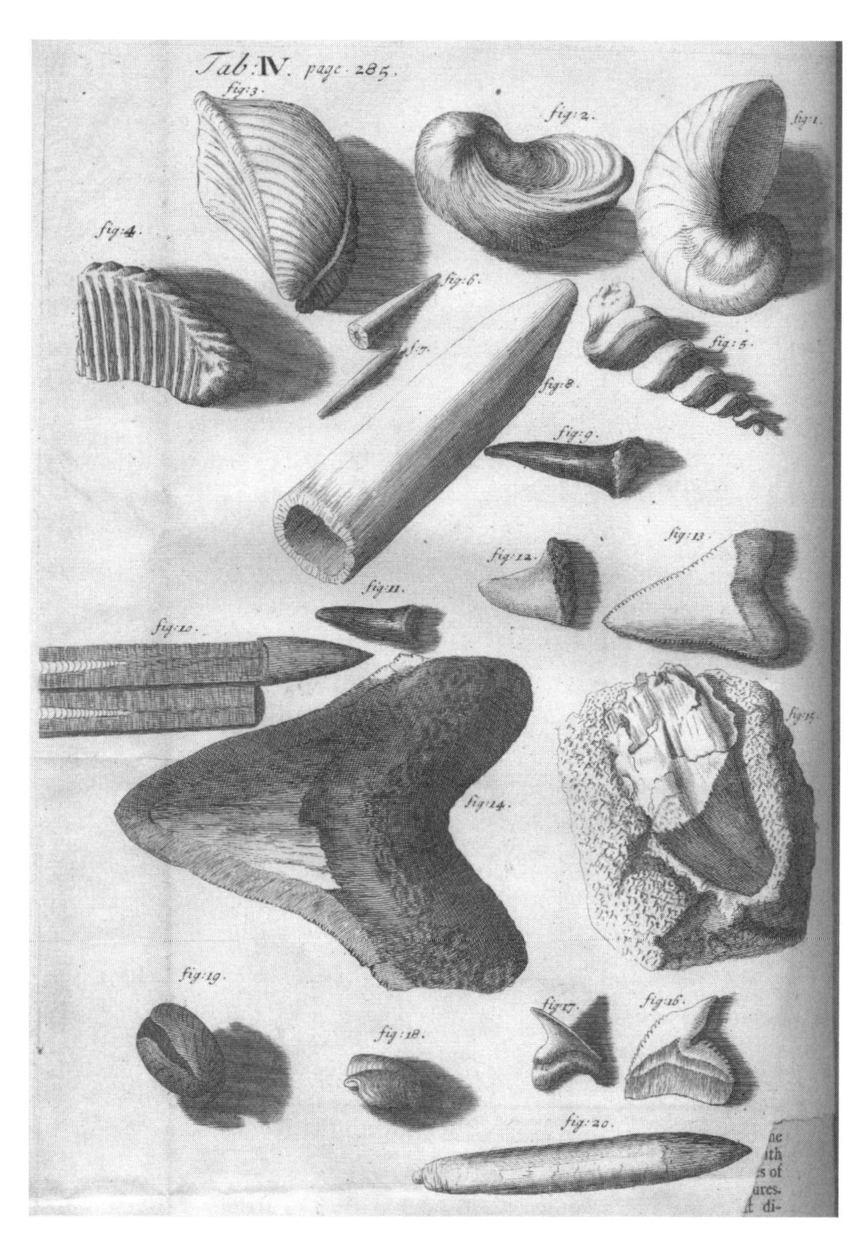

Fig. 53 Engravings of fossils from the Isle of Wight in Waller's edition of Hooke's *Posthumous Works*. Hooke almost certainly made the drawings on which these were based when visiting the Isle of Wight in 1666. (Museum of the History of Science, Oxford.)

This account gives us an intriguing glimpse of a man usually to be encountered in an urban setting, taking no more exercise than a brisk walk between coffee houses and his various places of work. Here he strikes out into a familiar seaside landscape with his walking-stick, clambering up the cliffs of his childhood, and scrambling among the rocks on the shore-line in search of fossil specimens.[55] Here too, Hooke's tone suggests a man at ease with his surroundings, in spite of the unfortunate circumstances which had necessitated his visit to the old family home. Still, in terms of Hooke's relentless pursuit of scientific data to inform his myriad intellectual interests, we have here one more example of the strain produced by competing claims on his time and energies.

Hooke was back in Epsom doing more well experiments (in the bitter cold) in January 1666. The candle experiments worked better, but Hooke does not record any experiments involving lowering a person to the bottom of the well.

The plague outbreak over, Hooke returned to his lodgings at Gresham College at the beginning of February, when he was probably back in London shortly before Wren (from Paris) and the court (from Oxford).[56]

The purpose of mapping these distinct, intersecting, and competing activities of Hooke's, during a relatively short period of intense experimental activity, has been to underline the extraordinary level of pressure under which Robert Hooke operated on a daily basis. Hooke had the capacity successfully to carry out intellectually and physically taxing work in several different fields simultaneously, sometimes leading the programme of work himself, sometimes collaborating closely with different individuals embarked on specific interests of their own. In spite of later claims that he was of a somewhat fragile disposition (based largely on a single suggestion that in adolescence paint fumes gave him headaches), we have evidence here of a robust constitution, and a remarkable ability to compartmentalize different activities in such a way that they did not interfere with one another.[57]

In the early years of the Royal Society, the sequence of activities we have looked at so far provides first-hand evidence that Hooke was in great demand in scientific and intellectual circles, expected to carry out a daunting number of distinct programmes of practical activities, in a wide range of distinct disciplines, on behalf of an equally wide range of friends, clients, employers, and former employers.[58] The pressures exerted on his time by his various esteemed colleagues established a habitual pattern for his life, which (I would argue) continued thereafter, until shortly before his death. He would move rapidly from commitment to commitment, from topic to topic, and from location to location (both within and beyond London).[59] He kept his many activities ruthlessly compartmentalized, right down to the fact that he moved from group to group of his acquaintances (each frequenting different coffee-house locations, or different districts of London), without allowing them to overlap. He documented transactions meticulously for each occupation—his accounts and reports on his London surveys, for instance, are exemplary for the period.[60] And he kept succinct notes of the progress of each of his many activities in his personal diary.[61]

For the remainder of this chapter I shall argue that it was as a direct consequence of this exacting occupational lifestyle that yet another area of Hooke's strenuous, systematic experimenting, sifting of data, and meticulous recording of outcomes took the form it did—his self-experimenting with some of the newest pharmaceuticals available to physicians and medical experimentalists. In this case, the evidence is to be found in Hooke's diaries, covering significant parts of the 1670s, 1680s, and 1690s.

Medical Case Notes

Incomplete diaries survive for two periods of Hooke's life.[62] The first, and most complete, covers the period 1672 to 1683.[63] The second begins shortly before the arrival in England of William of Orange, and the Glorious Revolution of November 1688, and continues until August 1693.[64] The earlier diary, in particular, closely corresponds to the kind of record needed to keep track of the detailed demands of the various distinct activities with which Hooke was involved (Fig. 54)—by the 1670s he had added

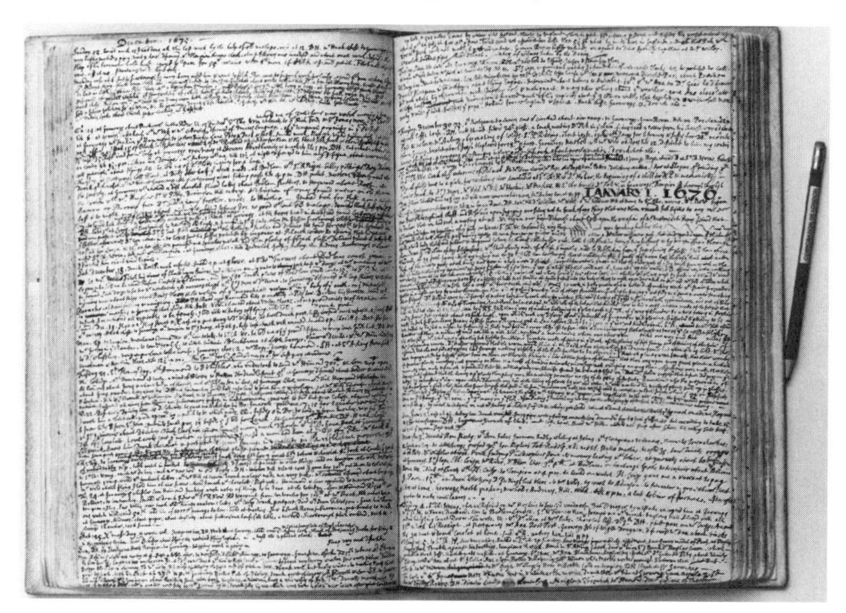

Fig. 54 A page from the manuscript of Hooke's diary entry for 1 January 1676, with sketches of wave and pulse forms of light. The wave and pulse hypotheses were proposed by Hooke and Wren to explain observations by Newton on refraction and colours. A few members were meeting for the first time as a private club within the Royal Society. (Guildhall Library, MS 1758.)

the gruelling work as London City Surveyor to his already busy schedule, and was the salaried 'first officer' in Wren's thriving architectural firm.[65]

Before focusing on selected episodes from Hooke's surviving diaries to develop further our sense of his packed, often conflicting, obligations and activities, it may be helpful to offer a typical series of entries, to give some idea of their meticulous, orderly diversity.[66] Here is a run of entries for the first five days of September 1672:

> (1) Drank Steel [therapeutic solution containing iron].[67] benummd my head, somewhat hotter all day. Eat milk. Saw Mr. Wild at Garways [coffee house]. Calculated length of glasses [lenses for telescopes/ microscopes]. Mr. Moor here. I invented an easy way for a musick cylinder with pewter tipes pinched between cylindrick rings. Slept ill all night and observed Mars with speculum [telescope], but not so good.
>
> (2) Grace [Hooke's 12-year-old niece and lodger] went to school after dinner. At Mr. Hauxes. Home. At Dr. Goddards tastd tinture of wormwood not spirit of wormwood [unfermented absinthe]. Made instrument for [viewing the] eclipse of starrs by the Moon. Eat raw milk, wrapd head warm and slept well after.
>
> (3) At Skinners Hall, Cotton. Guildhall, Youngs subpoena. Certificat at Fleet ditch next Boucheret 10sh. and 10sh. of Sir W. Humble and Lady Hoskins, Pauls churchyard, Mr. Godfrys. Controuler at Guildhall. Towne clarke and Mr. Rawlins about stable. Bought of Mr. Collins 5sh. farthings. Dr. Wrens, Garways, eat boyld milk. Slept pretty well, [orgasm]. took [ii oz.] of infusion of Crocus metall[icus] [iron oxide], vomited.
>
> (4) B. Bradshaw small pox. Purged 7 times [still the effect of the crocus metallicus]. eat dinner well. Disorderd somewhat by physick, urine had a cloudy sediment, but brake not. Slept pretty well.
>
> (5) At Guildhall, / Sir J. Laurence, Sir Th. Player. Home all afternoon. Mr. Haux [Haak]. Drank ale, eat eggs and milk, slept very little and very disturbed.[68]

These entries keep a continuous record of (at least) three distinct areas of Hooke's energetic activities: scientific experimentation (here largely astronomical, including both the manufacture of instruments and observation); city meetings, professional surveying activities, and other duties associated with the rebuilding of London after the Great Fire; and a regimen of diet and dosing, designed to address symptoms of personal ailments.[69] The first two tend to corroborate information, or supply addition detail, to materials to be found elsewhere—either in the Guildhall records, or among the Wren, Royal Society, and other professional papers. The third activity, however, is exclusive to the diary, although its detail corresponds closely to other equivalent evidence to be found in the writings and correspondence of those associated with Hooke.[70] Robert Boyle's correspondence, in particular, contains frequent references to pharmaceutical preparations requested by him, and supplied to him by physician friends.

Even from this short run of entries it will be apparent that Hooke was a chronic insomniac, and that whatever other medical symptoms he was attempting to deal with, his vigilance concerning his diet and dosing were tied up, in particular, with acute anxiety about getting a good night's sleep. We also get the sense—confirmed by the diary as a whole—that Hooke's physic-taking was regular and habitual; he dosed regularly to make himself feel better, rather than to treat individual, specific sicknesses.[71] Indeed, during the entire period of the diary, Hooke succumbed to no serious illness, instead recording, and treating, annoying symptoms such as giddiness, headaches, clouded vision, and palpitations, all of which may have been side-effects of the preparations he was consuming.[72]

The further symptoms for which Hooke dosed himself were ones which we might plausibly associate with the hectic intellectual and professional life scrutinized in some detail above: insomnia, headaches, dizziness, panic attacks, and problems with his eyes. His attempts at 'cures' for these persistent ailments followed several routes. Sometimes he consulted medical friends, and dosed himself with remedies they supplied. The 'steel' (solution of iron) which Hooke took at the beginning of September 1672, for instance, had been supplied to him by Dr Jonathan Goddard, Gresham Professor, Fellow of the Royal Society, colleague and companion, a month earlier. Before that, he had been taking a daily dose of a substance containing iron and mercury.[73] On another occasion he sampled 'tincture of wormwood' (absinthe before fermentation) with Dr Goddard.[74]

It was Goddard who supplied Hooke with his laudanum[75]—one of Hooke's habitual remedies for stress and sleeplessness:[76]

> [October 1672] (25) D[ined] H[ome]. Mr. Haux, Dr. Grew, Mr. Hill, Lord Brounker, Dr. Godderd, here, made [mercury] with a brass hole stand 3 inches and better. Dr. Godderd presented Mr. Gunters picture. Lord Brounker, Mr. Colwall and Dr. Godderd promisd theirs. The last time of meeting at my lodgings. Took conserves and flowers of [sulphur] after which I slept well, but had a bloody dysentry the next day (26) soe that I swouned and was violently griped, but I judge it did me good for my Rhume. About 6 at night I eat chicken broth and milk thicked

> with eggs and took conserve of Roses and Dr. Godderds Syrupe of poppys. Slept disturbedly but had no more griping. Mr. Haux, Mr. Blackburne, Mr. Mayer stayd with me in the afternoon.[77]

Inevitably, Hooke got conflicting advice from his medical friends. After a period of taking Dulwich water (a proprietary mineral water), he was advised by Dr Whistler that serious illnesses had been contracted by some of those drinking it, and that 'spirit of sal ammoniack [ammonium chloride]' was vastly preferable.[78] A couple of years later, when Hooke had largely moved on to other, newer pharmaceuticals like senna, Whistler's caution looked like wise advice when Hooke dosed himself (unsupervised) with Dulwich water, which dramatically disagreed with him:

> Monday 1 June, 1674.—Drank whey, took in Dulwich water, 3d. Eat bread and butter. Subpoenad by carpenter and Andrewes. Wrote this account. Was very ill that day. Was at view at Pudding Lane. The vomit did me much harm. At the pillar at Fish Street Hill. It was above ground 210 steps. DH. Received a letter from Sir J. Cutler.[79]

At times when he became more seriously debilitated by illness, Hooke consulted the same physicians more systematically, following closely the regimens they prescribed. In December 1672, between the death of Hooke's old mentor and close friend John Wilkins (Bishop of Chester, and the original founder of the Royal Society), Hooke became deeply depressed and troubled with persistent giddiness. He began by consulting three physician acquaintances:

> [8 December 1672] At home till 5. Well all day. taken at Garways with vertigo and vomiting. Dr. Ferwether, Dr. Godderd and Dr. King here. Took clyster and blister, wrought all night. Slept well. Mr. Whitchurch[80] here.[81]

Hooke's own view was that the giddiness was due to a combination of things he had eaten, and chill to the head caused by wearing a wig: 'going abroad found a great heavyness and guidiness in my eyes. . . . I guess of the guiddiness might proceed partly from drinking milk and posset drink and night and partly from coldness of perruck.' Ten days later, however, his symptoms were worse and he tried 'Dr. Godderd tincture of amber' (after taking which he got a good night's sleep, though, having inadvertently sat that evening by a cheering fire, he maintained that 'Fire made guiddiness worse'). On the night of 19 December Hooke 'Slept very little, was very giddy all next day'. Consulting Dr. Goddard again,

Goddard suggested it was trouble with his eyes that was causing the problem. Hooke 'tryd spectacles, found them help sight'; that night he slept better, with a little help from tobacco.[82]

Over the holiday period, Hooke resorted to a desperate series of treatments, for which he now paid a succession of physicians, though he continued to add remedies suggested more informally by his friends.[83] The impression gained from his diary entries is of a man weakened and increasingly depressed by his 'physic':

> (22) Mr. Gidly let me blood 7 ounces. Blood windy and melancholly, gave him 1/2 crown. Mr. Chamberlain here. the vertigo continued but upon snuffing ginger I was much relievd by blowing out of my nose a lump of thick gelly. Slept after it pretty well. went abroad (23) in the morn returnd home very guiddy. Refresht by eating Dinner, in the afternoon pretty well. Mr Lodowick, Lamot, chirugeon, etc. here. Physitian ordered 20sh. Consulted Dr. Godderd, he advisd amber and ale with sage and rosemary, bubbels, caraways and nutmeg steepd and scurvy grasse.
>
> (24) . . . DH. very ill and giddy in the afternoon but pretty well before. I took a clyster [enema] after which working but once I was very ill and giddy. Slept little all night save a little next morn. The worst night I ever yet had, melancholy and giddy, shooting in left side of my head above ear.
>
> (25) Christmas day. Slept from 7 to 10, rose pretty well, upon eating broth, very giddy. Mr. Godfry here but made noe effect. Eat plumb broth, went pretty well to bed but slept but little and mightily refresht upon cutting off my hair close to my head and supposed I had been perfectly cured but I was somewhat guiddy.
>
> (26) next day and tooke Dr. Godderds 3 pills which wrought 14 times towards latter end. I was again very giddy and more after eating, which continued till I had taken a nap for 1/2 howr about 5 when I was very melancholly but upon drinking ale strangly enlivernd and refresht after which I slept pretty well and pleasantly. Dreamt of riding and eating cream.
>
> (27) After I was up I was again guiddy and was so for most part of the day. Sir W. Jones Mr. Haux, Mr. Hill, Dr. Pope, Mr. Lamot, here. Borrowed Mr. Colwalls ale, agreed not, guiddy head, benummed and guiddy. made oyle of bitter almonds, put some in right ear. . . . Slept ill.
>
> (28) With Lamot at Lord Brounkers. Mr. Haux at home. At Guidlys I made an issue in my Pole. Dr. Chamberlaine was here and directed. He made it with caustick, I gave him 5sh., he did not scarrify.

(30) Slept somewhat better but sweat much, head somewhat easd but about noon I was again very giddy. . . . DH. Mr. Haux, chess. very guiddy and eye distorted. About 9 at night Gidly applyd 2 cupping to my shoulders and scarrifyd whence came about 5 [ounces] of very serous blood. My head eased a little and I slept pretty well the following night without sweat or heat.

(31) Rose about 10, but guiddy. Eye much distorted (Dremt of a medicine of garlick and the night before I drempt of riding and of eating cream with Capt. Grant) . . . Drank a little sack and spirit of amber going to bed. Slept very little.[84]

Hooke's state of mind was clearly not helped by the fact that he had little company over this period, attended no Christmas dinner, and indeed, participated in no holiday festivities of any kind. Wilkins's death, and the mourning period which followed, may have had something to do with this. Hooke did not recover his health until the middle of January, and there is no indication that any of his medication did more than alleviate his symptoms. (At the same time, they probably produced equally worrying side-effects.)

Hooke's treatment of this illness was in many ways uncharacteristically unsystematic, probably because he was feeling extremely sick and disorientated, rather than attending to persistent minor ailments.[85] In general, his self-dosing was carefully monitored, the details of dosage and outcome scrupulously recorded in his diary. In 1672, he kept a similarly precise record of the treatment administered to a close friend—none other than Bishop John Wilkins, whose subsequent death sent Hooke into depression that Christmas.

On 16 November 1672, Hooke recorded in his diary that he had learned in conversation with Boyle and Wren that Wilkins was gravely ill 'of the stone', as it was thought. A team of Royal Society medical men prescribed remedies, including Hooke's friend and personal physician, Dr Goddard:

(16) Blackfryers, Bridewell, Dr. Wren, Mr. Boyle, Cox. Lord Chester [Wilkins] desperately ill of the stone, stoppage of urine 6 dayes. oyster shells 4 red hot quenched in cyder a quart and drank, advisd by Glanvill. Another prescribed flegma acidum succini rectificatum cum sale tartari. Dr. Godderd advisd Blisters of cantharides applyd to the neck and feet or to the vains.[86]

The next day Hooke visited Wilkins and found him somewhat better, but a day later, when Hooke visited again, he had taken a turn for the worse, and early the following day a friend called at Hooke's residence at Gresham College to inform him that Wilkins had died. Hooke spent the remainder of the day in the company of other friends of Wilkins's, including Wren, and conversation turned repeatedly to alternative therapies which might have cured Wilkins of his kidney stones:

> (18) At Lord Chesters, he was desperately ill and his suppression continued. Bought a retort and separating glasse. Shortgrave, paid 7sh.
> (19) Mr. Lee here. Lord Bishop of Chester dyed about 9 in the morning of a suppression of urine. Dr. Wren here at Dionis Backchurch. Dind at the Bear in Birchen Lane with Dr. Wren, Controuler, Mr. Fitch. At Jonas Mores, sick, he was cured of a sciatica by fomenting the part for an hour with hot steames for one hour afterward chafing in oyles with a rubbing hand and heated firepans, which gave him a suddain ease. Sir Theodore Devaux told me of Sir Th. Meyerns cure of stone in kidneys by blowing up bladder with bellows etc.

On 20 November, however, Hooke was present at dinner with other Royal Society members when the doctor who had performed an autopsy on Wilkins's body arrived and reported that no sign of kidney stones had been found. (Like many other Fellows of the Royal Society, Wilkins had made his body available for 'experiment' after death, to determine the cause of his fatal illness):

> (20) DH. with Mr. Hill to Arundell House [Royal Society meeting], Experiment of fire and air, Flamsteed's letter. Supd at Kings head with Society. Dr. Needham brought in account of Lord Chesters having no stoppage in his uriters nor defect in his kidneys. There was only found 2 small stones in one kidney and some little gravell in one uriter but neither big enough to stop the water. Twas believd his opiates and some other medicines killd him, there being noe visible cause of his death, he died very quickly and with little pain, lament of all.[87]

Hooke and his medical colleagues administered highly toxic 'physic' to the body of their sick friend, and watched for improvement in the symptoms which did not come. Chemically, these acidic remedies could be thought of as combining in the blood to dissolve the chalky solids of the stone. One of the phenomena, the greatly magnified image of which Hooke inspected, and reproduced in his popular illustrated treatise on the microscope, *Micrographia* (1665), was his own clouded urine—urine which contained, in suspension, a chalky substance, or 'gravel' (Fig. 55). Hooke announced that close examination revealed this 'gravel' to be crystalline:

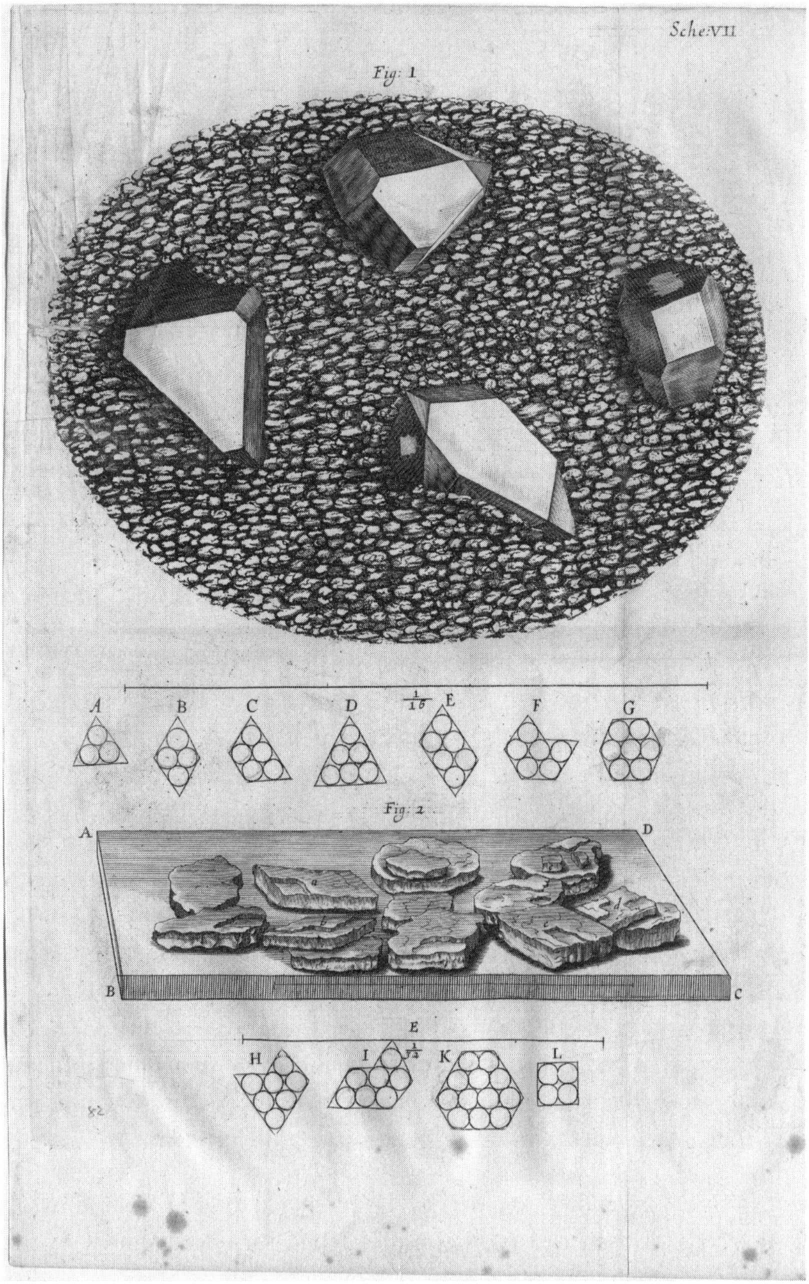

Fig. 55 Hooke's illustration (Fig. 2) of 'gravel' in urine, microscopically enlarged, from *Micrographia*. Having shown in the laboratory that these crystalline deposits dissolved when combined with various acids, Hooke proposed injecting a suitable solvent directly into the body to dissolve bladder-stones 'without hurting the Bladder' (that is, without surgical intervention). (Museum of the History of Science, Oxford.)

I have often observ'd the Sand or Gravel of Urine, which seems to be a tartareous substance, generated out of a Saline and a terrestrial substance crystalliz'd together, in the form of Tartar, sometimes sticking to the sydes of the Urinal, but for the most part sinking to the bottom, and there lying in the form of coorse common Sand; these, through the Microscope, appear to be a company of small bodies, partly transparent, and partly opacous, some White, some Yellow, some Red, others of more brown and duskie colours.

Hooke found that the crystals could be made to dissolve again in 'Oyl of Vitriol [concentrated sulphuric acid], Spirit of Urine [impure aqueous ammonia, containing ammonium carbonate], and several other Saline menstruums'.[88] Might not this offer a remedy for that most widespread of seventeenth-century complaints, 'the stone', he suggested? What was required was a solvent for the crystals harboured by the blood, which could be injected directly into the bladder:

> How great an advantage it would be to such as are troubled with the Stone, to find some menstruum that might dissolve them without hurting the Bladder, is easily imagin'd, since some injections made of such bodies might likewise dissolve the stone, which seems much of the same nature.[89]

In *Micrographia*, Hooke went on to suggest that Wren's method for injecting liquids into the veins of animals (which Hooke hinted had been more widely tried than had been reported, by Boyle among others) might be adapted to introduce solvents into the bloodstream of those suffering from a variety of common ailments:

> Certainly, if this Principle were well consider'd, there might, besides the further improving of Bathing and Syringing into the veins, be thought on several ways, whereby several obstinate distempers of a humane body, such as the Gout, Dropsie, Stone, &c. might be master'd and expell'd.[90]

Failure of the treatment in Wilkins's case is explained, for Hooke, by the fact that the diagnosis of Wilkins's inability to pass urine was incorrect: there was no stoppage in his urethra. Far from curing the condition, as Hooke laconically observes, it was the powerful toxins which probably proved fatal. The medical 'experiment' had failed, lending some irony to the report that as Wilkins lay dying, 'he said he was "ready for the Great Experiment" '.[91]

The Wilkins case shows us that dosing with what seem to us far-fetched (and potentially damaging) remedies was carefully consistent with current experimental scientific views, as described, for instance, in Hooke's analysis of chemical solvents for the chalky precipitate in blood.

Similar attempts at observing doses and outcomes in the physic-taking of distinguished friends and colleagues, and relating them as cause and effect, are recorded elsewhere in Hooke's diary—with sometimes tragic, sometimes less drastic results:

> 4 July 1673: 'This evening Sir R. Moray died suddenly being choked with flegme in indeavouring to vomit. he had Dind at Lord Chancellers and about an howr before his Death drank 2 glasses of cold water.'
>
> 30 October 1673: at Dr. Wrens he very sick with physick taking day before.[92]

Most of Hooke's diary entries concerning physic-taking, however, have to do with minutely recording his own intake of measured doses of medicaments and their outcomes, including both moderation of symptoms and production of side-effects. These are the laboratory notebooks of the habitual experimentalist, with himself as the experimental subject. Here are a few typical examples:

> 27 April 1674: Took Dr. Thomsons vomit [later: Slayer told me that Dr. Thomsons vomit was Sneider's *Antimonia Tartarizatis*]. It vomited twice. Purged 10 or 12 times. Read over discourse against Hevelius. At Garaways till 10. Made me sleep ill, and made my armes paralytick with a great noyse in my head. I had also some knawing at the bottom of my belly.[93]
>
> 30 July 1675: Took SSA [spirit of sal ammoniac] with small beer at Supper. Very feavorish all night but slept well. Strangely refresht in the morning by drinking SSA with small beer and sleeping after it, it purged me twice. [and the next morning] In a new world with new medicine.
>
> 1 August 1675: Took volatile Spirit of Wormwood which made me very sick and disturbed me all the night and purged me in the morning. Drank small beer and spirit of Sal-amoniack. I purged 5 or 6 times very easily upon Sunday morning. This is certainly a great Discovery in Physick. I hope that this will dissolve that viscous slime that hath soe much tormented me in my stomack and gutts. *Deus Prosperat.*[94]

Shortly after this, however, sal ammoniac ceased to produce detectable therapeutic results for Hooke (as habitual use accustomed his body to the toxin, so that it ceased to be effective), and he moved on to poison his system with new 'miracle' remedies.

Enhancing the Human Faculties?

The type of systematic self-experiment on which I want to concentrate for the remainder of this chapter is that carried out in relation to the healthy (or at least only slightly ailing) rather than the seriously sick body. It is self-experiment which explores a possible relationship between 'lightness' of spirit, alertness, intellectual clarity, and physic-taking. In a number of places Hooke links drug-taking with enhanced mental ability and clarity, as in one of the passages I cited above:

> Took Dr. Thomsons vomit [later: Slayer told me that Dr. Thomsons vomit was Sneider's *Antimonia Tartarizatis*]. It vomited twice. Purged 10 or 12 times. Read over discourse against Hevelius.

Hooke's diary gives some remarkable insights into this kind of medical investigation—one which of its very nature requires the experiment to be carried out on the experimenters themselves (to monitor inner states of mind known only to the patients).[95]

In the course of Hooke's illness of Christmas 1672, he noted a relationship between the outcome of his self-dosing and an accompanying mental state. The combination of a purge and a glass of ale produce sensations of freshness and alertness:

> Tooke Dr. Godderds 3 pills which wrought [purged] 14 times towards latter end. I was again very giddy and more after eating, which continued till I had taken a nap for 1/2 howr about 5 when I was very melancholly but upon drinking ale strangly enlivernd and refresht after which I slept pretty well and pleasantly. Dreamt of riding and eating cream.

Again, this matches a passage I have already quoted:

> Took SSA [spirit of sal ammoniac] with small beer at Supper. Very feavorish all night but slept well. Strangely refresht in the morning by drinking SSA with small beer and sleeping after it, it purged me twice.

And indeed, if we pay close attention, the scientific and dosing entries show a consistent pattern of association. This one (entered on 31 May 1674) indicates a perceived connection between the effects on the body of certain therapeutic substances and an increased clarity of the intellectual faculties:

> Took Childs vomit, Infusion Croc[us] Met[allicus]. i[oz]. Wrought pretty well. Refresht by it. D[ined] H[ome]. Made description of quadrant. Lost labour. Slept after dinner. A strange mist before my eye. Not abroad all day. Slept little at night. My fantcy very cleer. Meditated about clepsydra, quadrant, scotoscopes, &c.[96]

A dose of a proprietary emetic and a mildly toxic purgative ('crocus metallicus' is a preparation of iron oxide) produce clouded vision, wakefulness, and clarity of 'fancy'—thought or imagination. As a result, Hooke has particularly vivid and interesting thoughts about water-clocks, navigational instruments, and tools for microscopy. (His diary shows him repeating 'physic' which has thus enhanced his mental faculties.)

Again, Hooke's diary entries suggest that he treated his pharmaceutical experimenting as equivalent to other kinds of experiment he devised in his capacity as Curator of Experiments to the Royal Society, and alongside those he conducted in collaboration with Boyle, Wilkins, Petty, and Wren: '6 February 1674: At Spanish coffee house tryd new mettall with Antimony, iron and lead. At Shortgraves. Tryd reflex microscope.'[97] Hooke 'tried' a new toxic chemical compound as a medical remedy, and then 'tried' the reflex microscope. Both 'trials' involved skilled handling of the equipment or materials, and careful observation of outcomes and accuracy. The diary entry records two hands-on experiments in science and new technology, both meticulously noted in the day-by-day laboratory notebooks of the habitual experimentalist, with himself as the experimental subject. These are two domains within which Hooke expertly tests the latest thing in control and manipulation of the body and its senses. As Hooke makes clear in his preface to *Micrographia*, he believed that new precision scientific instruments like the telescope and microscope would give back to mankind the clarity of perception lost at the Fall:

> It is the great prerogative of Mankind above other Creatures, that we are not only able to behold the works of Nature . . . but we have also the power of considering, comparing, altering, assisting, and improving them to various uses. And as this is the peculiar priviledge of humane Nature in general, so is it capable of being so far advanced by the helps of Art, and Experience, as to make some Men excel others in their Observations, and Deductions, almost as much as they do Beasts. By the addition of such artificial Instruments and methods, there may be, in some manner, a reparation made for the mischiefs, and imperfection, mankind has drawn upon it self. [98]

By that double use of 'trial'—Hooke's customary term for developing such experimentally based 'helps' for the senses—we have at least a suggestion here that chemical 'instruments' might enhance man's post-lapsarian mental capacities, just as the microscope and telescope add magnification to his vision.[99]

As we eavesdrop on Hooke's self-experimenting with toxic purgatives, the patterns of dosing himself closely resemble his experiments in other, more conventional, scientific domains. Substances used are administered with close attention to causal relations between measured doses, combinations of substances, and direct bodily effects. Accompanying mental states—'strangely refresht', 'Refresht by it', 'much disturbed my head', 'made me cheerful'—are carefully noted. We recall again Bishop John Wilkins's supposed deathbed comment that he was 'ready for the Great Experiment', following the failed lesser experiments of administering solvents to disperse his 'stone', and that he was in 'little pain'.

It is also worth bearing in mind that a regimen of systematic self-dosing with purgatives was habit-forming. This was both because the body adjusted to the remedies over time, necessitating additional (and larger) doses to keep a state of equilibrium, or some kind of state of 'health', and because vomiting or excreting were commonly preceded by a state of elation (associated with the chemical side-effects of the drug)—leading Hooke to repeat the performance for the emotional 'lift' it gave his often depressed spirits.[100] On 16 February, 1673, for example, Hooke took Andrews cordial, which purged him, and, he also noted, in combination with tobacco, made him 'cheerful':

> [It] brought much slime out of the guts and made me cheerful. Eat dinner with good stomach and pannado at night but drinking posset upon it put me into a feverish sweat which made me sleep very unquiet and much disturbed my head and stomach. Taking sneezing tobacco about 3 in the morn clear my head much and made me cheerful afterwards I slept about 2 hours, but my head was disturbed when I waked.[101]

Only very rarely, for instance, on 3 August, 1673, a Sunday, does Hooke record that he 'took no physic'.[102]

On the basis of the evidence in Hooke's diary, one might, I think, suggest that here we have a regimen (a meticulously planned and observed programme of ingestion of food and drugs, over extended periods) which serves two functions simultaneously: it regulates bodily functions (meanwhile producing side-effects like numbness and blurred vision, which are then treated with further substances); and it excites the mental faculties, producing a clarity which is conducive to slightly fevered intellectual activity of the kind Hooke needed in order to cope with his chronic burden of overwork and the competing demands of clients and employers. It formed part of a serious programme for extending the reach of the human mind, and man's ability to control nature, alongside more conventional experimental programmes, such as those involving the microscope and the telescope.

Succumbing to Self-Experiment

The most damaging conflict of interests in Hooke's crowded and demanding life turns out, in the end, to have been his assiduous therapeutic self-dosing. On the face of it, this self-experimenting constituted the most systematic and sustained programme of experimental science in Hooke's varied repertoire—conducted more regularly and conscientiously even than his astronomical observations with long telescopes at Gresham College. Knowing what we now do about addiction, however, and about the accumulation of toxic substances in the system (imbibed over long periods, in small doses), it is probable that Hooke's regimen was ultimately fatal, and that even before it killed him, the side-effects of his medicines (clouded vision, giddiness, lassitude, melancholy) proved damaging and disabling.

By the age of 65 Hooke was a physical wreck, emaciated, and haggard; he was irritable and prone to paranoia; he rarely any longer bothered to attempt a proper night's sleep. In his late-twenties and thirties, though, during the period we have been looking at here, Hooke had been a man in his prime, something of a dandy, and a popular man-about-town. His self-dosing with pharmaceuticals helped him sustain the kind of punishing, over-demanding, metropolitan professional life we associate with the twenty-first century, not the seventeenth. His experimenting with mind-enhancing preparations may also have established a style for intellectual activity which has lasted down to our own times—not for Hooke the relaxed style of the genteel, virtuoso, natural philosopher. He operated in a permanent state of high tension, on the edge, wary and wakeful, under the permanent influence of stimulants.

Apart from the harm Hooke did to himself during his lifetime, his physic-taking habit certainly contributed to his long-lasting negative reputation after death. In his latter years he turned increasingly in on himself, cutting himself off from the broad and varied support network of friends who had fuelled his early career intellectually, and sustained his position socially. One of the themes taken up by each of the contributors to the present volume in turn has been that Hooke in his prime—during the period discussed at the beginning of this chapter—was popular, gregarious, and good-natured. His posthumous reputation, however, is that of a difficult, reclusive man, suspicious by nature and secretive in his behaviour.

Hooke died on 3 March, 1703. Although he had stepped down from his official responsibilities at the Royal Society in the late 1690s, after several bouts of ill health, he had carried on with a more restricted scientific social life and assorted intellectual activities until at least 1699.[103] His funeral was an appropriately grand occasion, and a fitting finale for one of the founding figures of London Restoration science. As his biographer, and the friend of his later years, Richard Waller wrote: 'His Corps was decently and handsomely interr'd in the Church of St. Hellen in London, all the Members of the Royal Society then in Town attending his Body to the Grave, paying the Respect due to his extraordinary Merit.'[104]

His last months, however, had been ones of ghastly deterioration, a literal rotting of his body, leading to a humiliating helplessness and a miserable passing away. Something of the stages of Hooke's physical decline in his final illness, and the way this impacted—disastrously and permanently—on his lasting reputation can be reconstructed from glimpses in a court case, the documents concerning which have recently come to light in the Chancery Court Records in London.[105]

As is made clear from what happened to Hooke's estate, there was really no one of substance and influence left to see to his posthumous reputation. In the absence of close heirs (since Hooke had never married, and his beloved niece Grace had died young) his estate went to two Isle of Wight cousins on his mother's side of the family, Elizabeth Stephens and Anne Hollis. They inherited

> great quantities of ready money in Gold and Silver and other medalls and a considerable library of Books and Jewells Plate Gold and Silver Rings Pictures Ornaments Statues Naturall and other Rarities and Curiosities of Great Value Household Stuff and Implements of household linnen woollen and apparrell and other Goods Chattells and Effects amounting in the whole to the value of Eighteen Thousand pounds and upward.[106]

Such, at least, is the claim made by a further claimant on Hooke's will, who appeared on the scene four years later. In 1707 another cousin of Hooke's, Thomas Giles, twin brother of Robert Giles, who had been Hooke's tenant on the Isle of Wight, and whose son Tom had, until his untimely death, been part of Hooke's domestic household in London, sent letters to the Court of Chancery in London, claiming a share in the Hooke inheritance. Since Thomas Giles was a prominent member of the Isle of Wight colony in Virginia, separated from his family by the Atlantic Ocean, it was understandable (he submitted) that he had not heard of Hooke's death until several years after it took place.

> After the decease of his said father Thomas Giles and many years before the said intestate [Hooke] died he went beyond the seas and was at the time of the said Intestates death residing and settled in Virginia in the West Indies and being [absent] at so great a distance did not hear of the said Intestates death till a long time after he dyed but for the unjust purposes herein after charged and set forth the said [Anne Hollis at the instigation or by or with the advice of Lewis Stephens of the parish of St James's Westm<inster>] and Elizabeth his wife industriously concealed the same from your Orator and taking advantage of your Orators ignorance thereof and of his absence and swore that the said Elizabeth was the only surviving [next of kin to the said Intestate of his fathers side and that she and the aforementioned] the only next of kin or nearest relation of the said Intestate Robert Hooke and that Your Orator was dead.[107]

In the course of Thomas Giles's complaint he describes these other members of the family dividing up Hooke's effects, with complete disregard for the great man or his memory. Indeed, Giles alleges that this process began even before Hooke's death:

> The said Confederates some or one of them did after the said Intestates death and during the last sicknings [his last illness] possess themselves or him or herself of all or the greatest part of the said Intestates Deedes Evidences Writings and securities for moneys Goods Plate Jewells Ready Moneys Effects and ch[arges?] on all Estate to such value and amount as herein before mentioned or thereabouts.

On the basis of further depositions lodged in the Giles Chancery case, it seems that Stephens and Hollis were actually *in situ* even before Hooke died, keeping a close eye on their prospective inheritance. We learn that as Hooke's health failed, and he began to need nursing attention at home, a coffee-house friend and instrument-maker, Reeve Williams, suggested that Hooke write to relatives on the Isle of Wight, appealing to them on grounds of kinship to come to his aid in his hour of need.

> These Defendants have Heard that the said Robert Hooke about Four or Five months before he dyed being in company with Reeve Williams one other of the Defendants . . . was complaining [that he lacked] a Servant Maid or to that effect and that the said Reeve Williams thereupon told the said Robert Hooke that he had heard the Complainants said late wife [Elizabeth Stephens] was some ways related to him and that she had a daughter married to one Joseph Dillon chairmaker and persuaded him to take one of them into his service.[108]

Hooke duly summoned Elizabeth Stephens, who brought with her Anne Hollis (also supposedly a relative) to assist her, since she herself was getting on in years: 'in regard your Orators wife [the deposition is by Lewis Stephens] was very aged herself and infirm and not of strength and ability of body to attend [on] and look after him'. Hooke apparently came to regret the decision (for the same deponent reports that 'the said Robert Hooke did then deny that the Complainants said late wife was anyways related to him'). His misgivings may have arisen from the fact that Stephens and Hollis apparently supplanted Hooke's close friends, who had until that point tended him lovingly themselves and, above all, protected his interests.

According to the testimony of the depositions in the Giles case, documents and valuables were systematically removed from Hooke's possession during his final illness. His properties on the Isle of Wight do indeed seem to have been quietly transferred into the names of others before Hooke's death. No wonder no trace remains of Hooke's many scientific instruments, nor, of course, of the portrait (or portraits) which apparently once existed. As Hooke lay dying, the relatives who tended him stripped him of his assets, pocketed his valuables, and took over the leases on his properties—his friends apparently being helpless to interfere.[109]

By the time of his death, Hooke had no close family to speak up for him, honour his memory, or protect his reputation. His appalling state of bodily health—the legacy of years of overwork and substance abuse—had, inevitably, largely distanced the once gregarious virtuoso from all but his most loyal friends.

Perhaps, as has often been claimed, it was Sir Isaac Newton, pursuing his grudge against Hooke beyond the grave, who denied his Royal Society colleague that posthumous fame he had deservedly earned in his lifetime. Under the circumstances, however, no such obsessive vindictiveness is needed to account for the eclipse of Hooke's reputation following his death. His vanishing from public view may simply have been the result of the circumstances surrounding his death, combined with the absence of a living, loving family to celebrate his achievements to the generation that followed.

Notes

1. Waller, 1705, xxvii.
2. Hunter, 1994c, 160.
3. For the issuing of the successive Charters of the Royal Society in 1662 and 1663 see Hunter 1989, 18–21.

4. For a full discussion of Hooke's various employments see Michael Cooper's section above.
5. In Oxford Hooke may have sometimes divided his working time between Boyle's laboratory and Dr Thomas Willis's dissecting room. See Frank 1980, 165.
6. Hooke lived at Lady Ranelagh's in Pall Mall until he was allocated accommodation at Gresham College in August 1664: 'I have also, since my settling at Gresham college, which has been now full five weeks, constantly observed the baroscopical index' (Hooke to Boyle, 6 October 1664, Boyle 2001, ii, 343). In March 1665 Hooke became Professor of Geometry at Gresham, which independently entitled him to rooms there, securing him accommodation for life. As well as designing and operating Boyle's famous air-pump, Hooke assisted Boyle with other technically complicated experiments, including experiments with mercury barometers, and with capillary action—the latter recorded in his earliest publication (1661). These latter experiments and observations were recapitulated in Royal Society discussions, and again in Boyle's correspondence with Huygens, and were finally published by Hooke under his own name (Hooke 1665). On top of his 'operator' activities, Hooke acted as a high-level scientific amanuensis, closely involved with seeing Boyle's works through the London press, and in particular, supplying his own diagrams for the engraved plates. See, e.g., Hooke to Boyle 24 November 1664: 'I have likewise procured out of Mr. Oldenburg's hands some of the first sheets; and shall delineate as many of the instruments you mention, as I shall find convenient, or (if it be not too great a trouble to you) as you shall please to direct. I think it will be requisite also, because your descriptions will not refer to the particular figures and parts of them by the help of letters; that therefore it would not be amiss if I add two or three words of explication of each figure, much after the same manner, as the affections of the prism are noted in your book of Colours. The figures I think need not be large, and therefore it will be best to put them all into one copper plate; and so print them, that they may be folded into, or displayed out of the book, as occasion serves' (Boyle 2001, ii, 412). The book in question is Boyle's *Cold.* None of these are 'menial' activities (compare the correspondence between Boyle and Robert Sharrock, who was doing the same job vis-à-vis Boyle's printed works in Oxford, e.g., Boyle 2001, ii, 293–4). Sharrock and Hooke clearly collaborated in some of this editorial work, see e.g., Sharrock to Boyle, 8 November 1660: 'I have according to my promise now Sent You up the translation of the first sheet [of the translation of Boyle's Spring of the Air into Latin] but I thincke it convenient that it bee transcrib'd before it bee return'd back for fear of miscarriage, Mr Whitaker, or Mr Hooke can read my Hand' (Boyle 2001, i, 436).
7. Boyle 2001, ii, 81 (my paragraphing). Oldenburg's insistence on the indispensable nature of Hooke's attendance as Curator of Experiments is at odds with Stephen Pumfrey's contention that during this period Hooke was merely one among a number of 'curators' at the Society. See Pumfrey 1991.
8. Boyle 2001, ii, 81–2.
9. Oldenburg to Boyle, 10 June 1663 (Boyle 2001, ii, 85–6).
10. Oldenburg to Boyle, 10 June 1663 (Boyle 2001, ii, 87).
11. Evelyn 1955, iii, 318.
12. In 1671 Hooke constructed an air-pump receptacle large enough to sit in, and occupied it himself while an operator removed a substantial portion of the enclosed air. He reported pain in his ears and giddiness. See 'Espinasse 1956, 51–2.
13. He was evidently assisting Boyle with astronomical observations. See Oldenburg to Boyle, 2 July 1663: 'The 8 or 10 dayes, which you allowed yourself in your letter for ‹the residu of your stay at› Lees, being now elapsed; yet the mixture of my fear, least the influence of that constellation, you then named, should arrest you longer, where you are, I thought, I would try, whether the operation of the starrs (since they are held but to incline) might ‹not› be diverted by the power of a Royal Society.' Oldenburg to Boyle, 10 June 1663 (Boyle 2001, ii, 95).

14. 'An experiment was tryed in the Compressing Engine, but again without successe,' Oldenburg to Boyle, 10 June 1663 (Boyle 2001, ii, 89). Actually, Hooke had not managed to get the air-pump working satisfactorily at the meeting he had been allowed to stay the additional days to attend—one more reason why Oldenburg was eager to get him back to London as quickly as possible.
15. Oldenburg to Boyle, 10 June 1663 (Boyle 2001, ii, 95). Boyle had still not returned to London on 18 July.
16. Bennett 1982, 51. Hooke's map of the Pleiades (a constellation within Taurus) is Hooke 1665, 241–2.
17. See Jardine 2002, ch. 5.
18. For Wren's to-ing and fro-ing see, e.g., Oldenburg to Boyle, 22 September 1664: 'This I intreat you, Sir, be pleased to communicate with my humble service to Dr Wallis, I name not Dr Wren, because he was present, when the letter itselfe was read to the Society' (Boyle 2001, ii, 329); Hooke to Boyle, 24 November 1664: 'Having received the honour of your commands by Dr. Wren, for procuring an instrument for refraction, whereby, as he told me, you designed to try the refraction of the humours of the eye; I did that very afternoon bespeak it; and I hope within a few days it will be ready to be conveyed to you' (Boyle 2001, ii, 412). The flow of information between Hooke and Boyle was public knowledge—see, e.g., Oldenburg to Boyle, 20 October 1664: 'I suppose, Mr Hook will himself answer your quere, which I acquainted him with, concerning the same' (Boyle 2001, ii, 357). Hooke had indeed answered Boyle's query in his letter of 21 October, from which it is clear that Wren is back in Oxford: 'I have not as yet any time to spend on these things, and therefore should be very glad, if yourself, or Dr. Wallis, or Dr. Wren, would examine what might be done in that kind; and what observations shall be further made, I shall most faithfully give an account of' (Boyle 2001, ii, 363).
19. See Jardine 2002, ch. 5.
20. 'At the time of writing Wren planned to return in December: "My Lord *Berkley* returns to *England* at *Christmas*, when I propose to take the Opportunity of his Company." This was George Berkeley, who had, while he himself was in Paris with Henrietta Maria, lent his house at Durdans, Epsom, to Wilkins, Hooke, and others, as a refuge during the outbreak of plague in summer 1665. On 30 November, Edward Browne reported in a letter that "Dr. Wren is at Lord Barclays" (Wren Society 1924–43, xviii, 178). However, Wren stayed in France until the end of February' (Jardine 2002, pp. 239–47).
21. Birch 1756–7, ii, 30.
22. Birch 1756–7, ii, 41.
23. Birch 1756–7, ii, 60.
24. Birch 1756–7, ii, 66.
25. Birch 1756–7, ii, 74.
26. Financial difficulties dogged Charles II's administration throughout his reign. At the beginning of 1672 the Crown effectively declared itself bankrupt with the 'Stop of the Exchequer', under which all repayments on its debts were suspended.
27. Jardine 1999, 149–53. For the full story of Hooke's complicated position in the longitude time-keeper patent race, see Wright 1989.
28. It may have been while they were both at Durdans that Hooke gave Wilkins an early version of his spring-balance watch as a gift.
29. In July Parliament was prorogued, and the Exchequer transferred to Nonsuch Palace at Epsom. Petty was presumably required to be in attendance there. See Pepys 1972–86, vi, 187–8.
30. Boyle 2001, ii, 493. Hooke goes on: 'I very much fear also, that we shall be forced against our wills to stay there long enough to try experiments of Cold [i.e., into winter], though I have some thoughts of removing to another place farther from London, where I have designed to try a larger catalogue of experiments, such as one cannot

every where meet with an opportunity of doing; but the country people are now so exceeding timorous, that they will not admit any, unless one have been a considerable time absent from London. I was this day informed by one, that received a letter thence, that the plague rages so extremely in Southampton, that sometimes there die thirty in a night; and that has made Portsmouth, and the isle of Wight so fearful, that they will suffer none to enter' (Boyle 2001, ii, 493). Hooke's mother had died in June, and Hooke is presumably here recording his inability to get home to the Isle of Wight to settle his family affairs: '1665 June 12 Burial, Newport. Mrs. Cisesly Hooke, widow' (Newport Parish Register).

31. Hunter suggests that Boyle left Oxford before 4 July, but the reference in Oldenburg's letter of 4 July actually only implies that Boyle is about to leave Oxford. Boyle 2001, ii, 488.
32. Boyle 2001, ii, 495. In the 'Publishers Advertisement to the Reader' in the published *Hydrostatical Paradoxes, Made out by New Experiments, (For the most part Physical and Easie.)* (1666) (Boyle 1999–2000, v, 189–279), the difficulty in obtaining suitable glassware for Boyle's experiments, during the period when he was out of London (including that at Epsom) is again alluded to:

> He resolv'd to repeat divers Experiments and Observations, that he might set down their Phaenomena whil'st they were fresh in his Memory, if not objects of his sense. But though, when he Writ the following Preface, he did it upon a probably supposition, that he should seasonably be able to repeat the intended Tryals, yet his Expectation was sadly disappointed by that heavy, as well as just, Visitation of the Plague which happened at *London* whil'st the Author was in the Country: and which much earlier then was apprehended, began to make havock of the People, at so sad a rate that not only the Glassmen there were scatter'd, and had, as they themselves advertis'd him, put out their Fires, but also Carriers, and other ways of Commerce (save by the Post) were strictly prohibited betwixt the parts he resided in and *London*; which yet was the only place in England whence he could furnish himself with peculiarly shap'd Flasses, and other Mechanical Implements requisite to his purposes. (Boyle 1999–2000, v, 191.)

33. Boyle 2001, ii, 535. This experiment does not appear in *Hydrostatical Paradoxes* (1666).
34. On 6 August, Boyle's sister, Lady Ranelagh (herself at Leese Priory with her sister) expressed alarm that Boyle intended to travel from Epsom to Oxford; in fact he had already set out, since he wrote to Oldenburg from there on the same date. Boyle 2001, ii, 503, 504.
35. 7 August 1665; Evelyn 1955, iii, 416.
36. Boyle 2001, ii, 513: 'I am still at Durdens, my lord Berkley's house near Epsom, where Dr. W[ilkins] only remains, Sir W. P. being gone to Salisbury.'
37. Birch 1756–7, ii, 30, 31–2, 40, and 48.
38. 'When Hevelius claimed that telescopic sights had never been tried on large instruments, Hooke replied that he had used several, "and particularly one of Sr. Christopher Wren's invention, furnished with two Perspective Sights of 6 foot long each, which I made use of for examining the motions of the Comet, in the year 1665." [Hooke 1679, 77] This was Wren's "double telescope" which Hooke describes elsewhere in the Animadversions (1674), as "two square Wooden Tubes, joyn'd together at the end next the Object by a Joynt of Brass, and the Angle made by opening of them, measured by a straight Rule." [Hooke 1679, 54; note also page 32] . . . Hooke gives a very detailed account in a lecture that was probably written in the period 1665–9 (see Waller, 1705, pp. 498–503)' (Bennett 1982, 42–3).
39. On astronomical instrument-based versus chronometer-based solutions to the longitude problem, see Andrewes 1996.

40. Boyle 2001, ii, 493–4.
41. Boyle 2001, ii, 512–13. Hooke goes on: 'The other [quadrant], which is yours [i.e., Boyle paid for it], I hope within a day or two to perfect it, so as to go much beyond the other for exactness.' He adds in a postscript that this is still to be paid for: 'There is something above thirty shillings due to Mr. Thompson. I have forgot the particular sum, but if I misremember not, it was thirty two shillings.'
42. Hooke had already shown his improved quadrant to the Society among the 'recess' work produced for the meeting on 21 March 1666: 'Mr. Hooke brought in a small new quadrant, which was to serve for accurately dividing degrees into minutes and seconds, and to perform the effect of a great one. It had an arm moving on it by means of a screw that lay on the circumference. But the complete description of it was referred to the inventor' (Birch 1756–7, ii, 69).
43. Hooke did hope to construct a purpose-built 'depth' or well at Greenwich, for observational purposes, alongside the Observatory in 1675. A well is shown on a plan of Greenwich, but apparently none was ever sunk. See Jardine 2002.
44. Hooke to Boyle, 15 August 1665, Boyle 2001, ii, 512.
45. Boyle 2001, ii, 537–8.
46. Boyle 2001, ii, 538.
47. Birch 1756–7, ii, 69–70. In addition to his experiments with 'depths' (in wells), Hooke mentions having carried out similar experiments with 'heights', 'both on the higher parts of Westminster Abbey, and also on the top of St. Paul's tower'.
48. Birch 1756–7, ii, 70–1.
49. Birch 1756–7, ii, 71.
50. For Hooke as instrument-maker see Jim Bennett's essay above.
51. 'I am going shortly for a little while into the Isle of Wight, and so perhaps may not till my return be able to make those trials; but I suppose the winter will not afford less instructive experiments than the other. And therefore what you shall please to suggest now will not come too late for winter experiments, especially if I can give order for making ready an apparatus for them before I take my journey' (Hooke to Boyle, 26 September 1665, Boyle 2001, ii, 538). On John Hooke, see the section of the Isle of Wight website dedicated to his life and family history: http://freespace.virgin.net/ric.martin/vectis/hookeweb/intro.htm.
52. Hooke means during the extended summer recess prompted by the plague outbreak in London, during which this trip was made.
53. Drake 1996, 176.
54. Drake 1996, 181. Elsewhere in the Discourse, Hooke specifies the location of his fossil-hunting even more precisely:

    > To this I shall add an Observation of my own nearer Home, which other possibly may have the opportunity of seeing, and that was at the West end of the Isle of *Wight*, in a Cliff lying within the *Needles* almost opposite to *Hurst-Castle*, it is an Earthy sort of Cliff made up of several sorts of Layers, of Clays, Sands, Gravels and Loames one upon the other. Somewhat above the middle of this Cliff, which I judge in some parts may be about two [word missing?] Foot high, I found one of the said Layers to be of a perfect Sea Sand filled with a great variety of Shells, such as Oysters, Limpits, and several sorts of Periwinkles, of which kind I dug out many and brought them with me, and found them to be of the same kind with those which were very plentifully to be found upon the Shore beneath, now cast out of the Sea. (Drake 1996, 232.)

55. For an excellent description of the geology of the Isle of Wight, and its influence on Hooke's lifelong interest in geology see Drake 1996, ch. 2.

56. Hooke to Boyle, 3 February 1666:

> The weather was so very cold, when we made these experiments, that made us hasten then so much the more; and I have not since had an opportunity to repeat them, though, God willing, I intend to make many other of the like kind, either there or elsewhere, some time this summer; and I have great hopes of having an opportunity of examining both greater depths and much greater heights, in some of our English mines, and some of the mountains in *Wales*, which, with some other good company, I design to visit this next summer. (Boyle 2001, iii, 49.)

57. On Hooke's childhood constitution see Cooper, above.
58. It has become customary—though with no particular basis in historical fact—to characterize Hooke as a menial, put-upon scientific servant, pulled hither and thither by the competing demands of a number of elite employers. See, e.g., Pumfrey 1991, Shapin 1989. There is no doubt that Boyle and Hooke held each other in mutual regard: see, for instance, Hooke to Boyle, 8 September 1664, 'Most Honoured Sir, I must in the first place return you my most humble acknowledgement for the honour and favour you have been pleased to oblige me with in your letter, which, to my power, I shall ever be ready to express my sense of' (Boyle 2001, ii, 315).
59. After the Great Fire in September 1666, Hooke had the added (onerous) commitments of Surveyor for the City of London, and delegated responsibilities for rebuilding London, within the Wren architectural practice. See Jardine 2002, and Cooper, above.
60. See Cooper 1997, 1998–9. For more information on Hooke's professional practice and documentation, see Cooper's essay above.
61. There is some irony in the fact that the one set of experimental results Hooke apparently did not document with sufficient care were those concerning his priority in inventing and developing the balance-spring mechanism for a reliable pocket-watch. During precisely the period when he was at Epsom and Oldenburg remained in London, Oldenburg and Moray were conducting a fraught correspondence with Christiaan Huygens (in Paris and The Hague) on this subject. On 8 January 1666, Moray wrote to Oldenburg: 'Hook concealed his invention about Watches too long: pray tell him not to do so with what other things hee hath of that kind. hee hath seen the folly & inconvenience of it' (Oldenburg 1965–86, iii, 9). It may, in the end, have been Wren who 'leaked' Hooke's balance-spring mechanism to the French savants. See Auzout to Oldenburg, 4 April 1666: 'Joubliois de vous dire que quelques uns travailloient icy a rendre les montres meilleurs aussibien que vous aves fait en Angleterre, mais que lon non voioit aucun succes. il ya apparance que nous devons beaucoup espere des meditations de M. Hook sur ce suiet car M. Wren ma dit icy quil en avoit desira reduit quelque chose en pratique. ce cera une belle Invention' (Oldenburg 1965–86, iii, 83). Wren was in France until early March.
62. Gunther argues that there is a third diary fragment, covering the period October 1681 to September 1683, in the British Library, MS Sloane 1039 (Gunther, 1930b, 577). These appear to be little more than jottings.
63. This diary is in the London Guildhall Library, and is transcribed in Hooke 1935.
64. BL MS Sloane 4024. This diary is partially transcribed in Gunther 1935, 69–265.
65. For the full range of Hooke's professional commitments see Michael Cooper's essay above.
66. For another recent discussion of the medical entries in Hooke's diary see Mulligan 1996. Mulligan too argues that Hooke's medical diary entries are 'an integral part of his scientific vision', and that the diary should be read as recording 'a self that was as subject to scientific scrutiny as the rest of nature', aimed at producing a natural history 'with himself as the datum'. However, she sees the medical diary entries as designed to provide correctives for Hooke's observations and experiments elsewhere, rather than as experiments in themselves.

67. 11 August 1672: 'Steel drink from Dr. Goddard'. Hooke 1935, 4.
68. Hooke 1935, 4.
69. Although the most systematic recording of medical dosing and its outcomes is to be found in the earlier diary fragment, such entries also occur in the post-1688 diary, for example:

> Tue. 6 [November 1688]. Ticket for Com. Sewers Wensday 9. Very ill all day with a colick and stopage: slept ill. Water high colourd; fasted all day. W. 7. Very sore by the colllick. M[artha] made chocolot. Royal Society met: Herbert, Hill, Waller, Pitfeild, Lod. Visited Tison [Hooke's physician] severall times. Dutch sayd to be landed at Tor Bay. Malpighius letter read. Pitfield restored 2 vol. of Plutarck and Holl[ands] Pliny in English. I deliverd to Mr Hunt Waller his Algebra & 4 bookes of Mengoli, which he crost out of his book, HH severall times: here also Dr Tison. Beer & bread 17d. Th. 8. HH tea. Not yet well of Colick. Noe auditors in the morn. J. Mayor here. HH D[ined] with me. 3 in the afternoon, which HH answerd. Gof & Lod here: noe further news. Dr Tyson. (Gunther 1935, 71.)

70. For another article on Hooke's diary which focuses on his physic-taking see McCray Beier 1989.
71. McCray Beier, who compares Hooke's medicinal intake with that recorded in other contemporary diaries, describes Hooke as being of a particularly sickly disposition. It is hard to see how such a person could have kept up the gruelling agenda of work described above. On the other hand, Hooke's treatment of specified or unspecified 'symptoms' with medicinal intake was remarkable: 'For only eighteen months out of a total of 148 for the period between August, 1672 and December, 1680, he reported no symptoms at all. For 64 months, he mentioned illness or remedies on between one and eight occasions. And for 18 months, he mentioned these matters between nine and 19 times' (Beier 1989, 240).
72. See Beier 1989, 241: 'Despite what he and those around him perceived as his chronic ill-health, he suffered no serious illnesses at all in the period during which he kept his diary. His illnesses were annoying, but neither frightening nor life-threatening.'
73. 'August [1672]. (1) Drank [iron] and [mercury]. At Wapping with governors. Took beet, slept not well. Paid Coffin £2 3s. 4d. for shutts etc. (2) Drank [iron] and [mercury]. Dind at Swan, old fish. Capt. Clark. Committee at my chamber. (3) Drank [iron] and [mercury]. Dind at home. Scotland yard. Cox turning object glasses with flints. Mr. Gale. (4) Drank [iron] and [mercury]. DH. Eat coadling. was ill.' (Hooke 1935, 4).
74. Hooke 1935, 4: 'At Dr. Goddards tastd tinture of wormwood not spirit of wormwood.'
75. From Goddard's correspondence it appears that Boyle, in his turn, supplied Goddard.
76. For Hooke's regular taking of laudanum for insomnia see, e.g., Hooke 1935, 8–9:

> October [1672]. (1) At Bow with Kayus Sibber. I took spirit of urine and laudanum with milk for the three preceeding nights. Slept pretty / well.' '[6 October 1672] Slept very ill with great noyse in my head. (7) Drank Dulwich water. Chess with Mr. Haux. Garways. Bed. Dr. Bradford here with R. Waters and he promised to procure him his 150 within 3 weeks on condition he put the barrs into the windows before the Sunday following. Subpoenad by Cotton for the Skinners. Bought china. Received 1 pair of shoes and goloshoes from Herne for which and 1 pair of shoes I owe him 11s. 6d. Mr Shortgrave returned. (8) Drank Dulwich water which worked not well. [orgasm] Nell. . . . at door. Dr. Wren here, Mr. Fuller, Mr. Fitz. At home all the afternoon. Mr. Hauxes at night. took Dr. Goddard syrupe of Poppy, slept not.'

77. Hooke 1935, 11.
78. Hooke 1935, 8: 'At Dr. Tillotsons with Dr. Whistler. He said many had died of Fluxes and some of the Spotted feaver upon drinking Dulwich water. Mrs. Tillotson

recoverd of sowerness in her stomack upon taking Spirit of Sal armoniack. I found a great purging from my head after drinking Chester stale beer, and a very good stomack.'

79. Hooke 1935, 106.
80. Mr Whitchurch was one of the apothecaries who supplied Hooke with his pharmaceutical materials.
81. Wilkins's funeral took place on December 12: '(12) Slept well in my gown. Sawd wood, took down joyst. Kept in, rangd and catalogued Library. Lord Chester buried. (14) I drank a little milk going to bed this and the preceding night and slept ill after it, was feaverish and guiddy next day, but eat dinner with good stomack. Dean Tillotson brought me a mourning ring for Lord Chester.' (Hooke 1935, 16).
82. 'Slept pretty well, a breathing sweat. tobacco in nostril inclined to sleep.' (Hooke 1935, 17).
83. Once again we should note a certain tension between Hooke's own recourse to favoured remedies and the courses of treatment prescribed by his physicians. When Hooke's niece Grace was ill with smallpox in July 1679, she was attended by Mr. Whitchurch and Dr Mapletoft, who prescribed on the 26th. On the 27th Hooke administered 'Gascoyne's powder' to Grace (sent by her mother); when Whitchurch next visited he was 'angry', presumably at the taking of remedies other than those he had prescribed (Hooke 1935, 419).
84. Hooke 1935, 18.
85. Like Boyle, Hooke was obviously a considerable hypochondriac. On 15 April 1653 William Petty—who sometimes acted as Boyle's physician in Ireland—took Boyle to task for his hypochondria:

   > The next disease you labour under, is, your apprehension of many diseases, and a continual fear, that you are always inclining or falling into one or the other. . . . But I had rather put you in mind, that this distemper is incident to all that begin the study of diseases. Now it is possible that it hangs yet upon you, according to the opinion you may have of yourself rather then according to the knowledge that others have of your greater maturity in the faculty [of medicine]. But *ad rem*, Few terrible diseases have their pathognomonical signes; Few know those signes without repreated experiences of them, and that in others, rather then themselves: Moreover; The same inward causes produce different outward signes, and vice versa the same outward signes may proceed from different inward causes; and therefore those little rules of prognostication found in our books, need not always be so religiously beleived. Again 1000 accidents may prevent a growing disease itselfe, and as many, can blow away any suspicious signe thereof, for the vicissitude whereunto all things are subject, suffers nothing to rest long in the same condition; and it being no farther from Dublin to Corke, then from Corke to Dublin, why may not a man as easily recover of a disease without much care, as fall into it. My Cousen Highmores curious hand hath shewn you so much of the fabrick of mans body, that you cannot think, but that so complicate a peece as yourself, will be always at some little fault or other. But you ought no more to take, every such little struggling of nature for a signe of a formidable disease, then to fear that every little cloud portends a cataract or hericane.

   (Boyle 2001, i, 143–4.)
86. Hooke 1935, 13.
87. Hooke 1935, 13–14.
88. It was almost another 200 years before the medical profession came to understand the significance of the structure of the crystalline deposits in gout, and developed a genuine cure (see Porter and Rousseau 1998, ch. 5).

89. Hooke 1665, 81–2.
90. Hooke 1665, 144–5.
91. Crowther 1960, 47.
92. Hooke 1935, 49 and 67.
93. Hooke 1935, 99.
94. Hooke 1935, 172. Entries of this kind can be matched from Anthony Bacon's, Francis Bacon's, and Boyle's letters, and from Ward's, Pepys's, and Aubrey's memoranda.
95. Equivalent reports are to be found in the private jottings of other habitual users of chemical medicines, like Boyle and Sir Francis Bacon. See for example, the following, from Bacon, *Commentarius solutus*:

> After a maceration taken in the morning and working little I took a glister [enema] about 5 o'clock to draw it down better, which in the taking found my body full and being taken but temperate and kept half an hour wrought but slowly, neither did I find that lightness and cooling in my sides which many times I do, but soon after I found a symptom of melancholy such as long since with strangeness in beholding and darksomeness, offer to groan and sigh, whereupon finding a malign humour stirred I took three pills of aggregative corrected according to my last description, which wrought within two hours without griping or vomit and brought much of the humour sulphurous and fetid, then though my medicine was not fully settled I made a light supper without wine, and found myself light and at peace after it. I took a little of my troch[ises] of [sal] amon[iac] after supper and I took broth immediately after my pill [my emphases].

(See Jardine and Stewart 1998, 300–1). In the margin Bacon noted, in explanation: 'Note, there had been extreme heats for ten days before and I had taken little or no physic'—external factors, too, may contribute to the peace of mind and equanimity he achieves at the end of his 'regimen'.

96. Hooke 1935, 105.
97. Hooke 1935, 85.
98. Hooke 1665, preface, sig. a1.
99. Among Hooke's posthumously published papers is one on the therapeutic use of 'bangue' or marijuana, particularly for insomnia. See Hooke 1726, 1–3:

> An Account of the Plant, call'd Bangue, before the Royal Society, Dec. 18. 1689. It is a certain Plant which grows very common in India, and the Vertues, or Quality thereof, are there very well known; and the Use thereof (tho' the Effects are very strange, and, at first hearing, frightful enough) is very general and frequent; and the Person, from whom I receiv'd it, hath made very many Trials of it, on himself, with very good Effect. 'Tis call'd, by the Moors, Gange; by the Chingalese, Comsa, and by the Portingals, Bangue. The Dose of it is almost as much as may fill a common Tobacco-Pipe, the Leaves and Seeds being dried first, and pretty finely powdered. This Powder being chewed and swallowed, or washed down, by a small Cup of Water, doth, in a short Time, quite take away the Memory and Understanding; so that the Patient undestands not, nor remembereth any Thing that he seeth, heareth, or doth, in that Extasie, but becomes, as it were, a mere Natural, being unable to speak a Word of Sense; yet is he very merry, and laughs, and sings, and speaks Words without any Coherence, not knowing what he saith or doth; yet is he not giddy, or drunk, but walks and dances, and sheweth many odd Tricks; after a little Time he falls asleep, and sleepeth very soundly and quietly; and when he wakes, he finds himself mightily refresh'd, and exceeding hungry. And that which troubled his Stomach, or Head, before he took it, is perfectly carried off without leaving any ill Symptom, as Giddiness, Pain in the Head or Stomach,

or Defect of Memory of any Thing (besides of what happened) during the Time of its Operation. . . . The Plant is so like to Hemp in all its Parts, both Seed, Leaves, Stalk, and Flower, that it may be said to be only Indian Hemp. . . . [T]his I have here produced, is so well known and experimented by Thousands; and the Person that brought it [Robert Knox] has so often experimented it himself, that there is no Cause of Fear, tho' possibly there may be of Laughter. It may therefore, if it can be here produced, possibly prove as considerable a Medicine in Drugs, as any that is brought from the Indies; and may possibly be of considerable Use for Lunaticks, or for other Distempers of the Head and Stomach, for that it seemeth to put a Man into a Dream, or make him asleep, whilst yet he seems to be awake, but at last ends in a profound Sleep, which rectifies all; whereas Lunaticks are much in the same Estate, but cannot obtain that, which should, and in all Probability would, cure them, and that is a profound and quiet Sleep.

100. A number of the remedies used by Bacon, Hooke, and others are still listed among modern alternative medical therapies as 'combating melancholy' when administered in small doses. Wormwood (absinthe) is just one of the remedies which elates when imbibed, but over time acts on the nervous system, producing tremors and delusions.
101. Sometimes the remedy's intoxicating properties—its ability to give the patient a 'high'—were discovered accidentally. Hooke's friend and colleague Edmond Halley was so taken with the clearheadedness and calming effect of an over-large dose of opium he had inadvertently ingested (because it had been insufficiently mixed into a compound he had taken) that he read a paper on it to the Royal Society in January 1690, describing how 'instead of sleep, which he did design to procure by it, he lay waking all night, not as if disquiet with any thoughts but in a state of indolence, and perfectly at ease, in whatsoever posture he lay' (Halley 1937, 217).
102. See Beier 1989, 244.
103. See the volume of Hooke papers in the archives of the Royal Society, Classified Papers. XX, which contain works 'in progress', including work for the Society, down to May 1699—latterly reading and producing synopses of new scientific books, in increasingly tiny, obsessive writing.
104. Hooke 1705, xxvi.
105. Thomas Giles c. Joseph Dillon. See C9 194/40 and C24/1278.
106. Thomas Giles's complaint to the Court of Chancery, 12 February 1706: PRO C/9/194/40. The first response to this complaint by Joseph Dillon (dated 6 November 1707) includes a copy of the original inventory of decease which is in the PRO, and which is printed in Hunter and Schaffer 1989, 292–4. This records goods to a significantly lower total value—Giles would presumably have argued that it was taken after valuables had already been removed. I am grateful to Ellie Naughtie for her help in transcribing these documents.
107. Passages in square brackets are interpolated into the text, above the line, in another hand.
108. PRO C5 335/67—Stephens v. Hollis. Passages in square brackets have been added for clarification.
109. But for a suggestion of how Wren, Waller, and Hunt may have managed to acquire at least enough of Hooke's inheritance to endow the Royal Society with a 'Repository' or purpose-built museum, in accordance with Hooke's known last wishes, see Jardine 2002, ch. 8.

BIBLIOGRAPHY

FURTHER READING

Two older accounts of Hooke that remain worth reading are Margaret 'Espinasse's zestful account of his life and ideas, published by William Heinemann in 1956 (and reprinted by the University of California Press in 1962), and E. N. da C. Andrade's 1949 Wilkins Lecture to the Royal Society, published in *Proceedings of the Royal Society, Series A,* 201 (1950), 439–473 (also printed in *Series B,* 137, 153–187). A more recent overview is provided by the essays in the volume edited by Michael Hunter and Simon Schaffer, *Robert Hooke: New Studies* (Boydell Press, 1989), which stems from a conference on Hooke held in 1987. Beyond that, readers are recommended to sample some of Hooke's own writings, perhaps notably his most famous book, *Micrographia* (1665), which is available on CDRom (Octavo, 1998, available at http://www.octavo.com/collections/projects/hkemic/index.html). Further primary and secondary literature on different facets of Hooke's activities is cited in the notes to this book, and is comprehensively listed in the Bibliography that follows.

BIBLIOGRAPHY

Aarsleff, Hans (1980). John Wilkins. In *Dictionary of Scientific Biography,* ed. C. C. Gillispie, Charles Scribner's Sons, New York, 14: 361–381.

Adamson, Ian (1978). The Royal Society and Gresham College. *Notes and Records of the Royal Society* 33: 1–21.

Agassi, Joseph (1977). Who discovered Boyle's Law? *Studies in History and Philosophy of Science* 8: 189–250.

Andrade, E. N. da C. (1950). Wilkins Lecture: Robert Hooke. *Proceedings of the Royal Society, Series A*, 201: 439–473 (also printed in *Series B*, 137: 153–187).

Andrews, W. J. H. (1996). *The Quest for Longitude*. Harvard University Press, Cambridge, Mass.

Anon. (1842). Gallery of Illustrious Irishmen, No. XIII. Sir Thomas Molyneux, Bart., M. D., F. R. S. *Dublin University Magazine* 18: 305–327, 470–490, 604–619, 744–764.

Anstey, Peter (2002). Locke, Bacon and natural history. *Early Science and Medicine* 7: 65–92.

Aubrey, John (1898). *Brief Lives, Chiefly of Contemporaries*, ed. Andrew Clark. 2 vols. Clarendon Press, Oxford.

Aubrey, John (2000). *Brief Lives*, ed. John Buchanan-Brown. Penguin, London.

Bacon, Francis (2000). *The New Organon*, ed. Lisa Jardine and Michael Silverthorne. Cambridge University Press, Cambridge.

Barker, G. F. R. (1895). *Richard Busby D. D.* Lawrence & Bullen, London.

Batten, M. I. (1936–7). The architecture of Dr Robert Hooke, FRS. *Journal of the Walpole Society* 25: 83–113.

Bechler, Zev (1974). Newton's 1672 optical controversies: a study in the grammar of scientific dissent. In Yehuda Elkana (ed.), *The Interaction between Science and Philosophy*. Humanities Press, Atlantic Highlands, NJ, 115–142.

Beier, L. M. (1989). Experience and experiment: Robert Hooke, illness and medicine. In Hunter and Schaffer 1989, 235–252.

Bell, W. G. (1923). *The Great Fire in London in 1666*. Bodley Head, London.

Bennett, J. A. (1975). Hooke and Wren and the system of the world: some points towards an historical account. *British Journal for the History of Science* 8: 32–61.

Bennett, J. A. (1980). Robert Hooke as mechanic and natural philosopher. *Notes and Records of the Royal Society* 35: 33–48.

Bennett, J. A. (1982). *The Mathematical Science of Christopher Wren*. Cambridge University Press.

Bennett, J. A. (1989). Hooke's instruments for astronomy and navigation. In Hunter and Schaffer 1989, 21–32.

Bennett, J. A. (1997). Flamsteed's career in astronomy: nobility, morality and public utility. In Frances Willmoth (ed.), *Flamsteed's Stars: New Perspectives on the Life and Work of the First Astronomer Royal (1646–1719)*. Boydell Press, Woodbridge, 17–30.

Birch, Thomas (1756–7). *The History of the Royal Society of London*. 4 vols. London.

Birkett, Kirsten, and Oldroyd, David (1991). Robert Hooke, physico-mythology, knowledge of the world of the ancients and knowledge of the ancient world. In Stephen Gaukroger (ed.), *The Uses of Antiquity*. Kluwer, Dordrecht, 145–70.

Boyle, Robert (1999–2000). *Works*, ed. Michael Hunter and E. B. Davis. 14 vols. Pickering & Chatto, London.

Boyle, Robert (2001). *Correspondence*, ed. Michael Hunter, Antonio Clericuzio, and L. M. Principe. 6 vols. Pickering & Chatto, London.

Carleton, John (1938; rev. ed. 1965). *Westminster School: A History*. Rupert Hart-Davis, London.

Chartres, Richard, and Vermont, David (1998). *A Brief History of Gresham College 1597–1997*. Gresham College, London.

Centore, F. F. (1970). *Robert Hooke's Contributions to Mechanics*. Martinus Nijhoff, The Hague.

Cohen, I. B. (1964). Newton, Hooke and 'Boyle's Law' (discovered by Power and Towneley). *Nature* 204: 618–621.

Cohen, I. B. (ed.) (1978). *Isaac Newton's Papers & Letters on Natural Philosophy*. 2nd ed. Harvard University Press, Cambridge, Mass.

Colvin, Sir Howard (1995). *A Biographical Dictionary of British Architects 1600–1840*. 3rd ed. Yale University Press, New Haven and London.

Cooper, M. A. R. (1997, 1998a, b). Robert Hooke's work as Surveyor for the City of London in the aftermath of the Great Fire. *Notes and Records of the Royal Society of London* 51: 161–174, 52: 25–38, 205–220.

Cooper, M. A. R. (1999). *Robert Hooke, City Surveyor*. Unpublished PhD thesis, City University, London.

Crowther, J. G. (1960). *Founders of British Science: John Wilkins, Robert Boyle, John Ray, Christopher Wren, Robert Hooke, Isaac Newton*. Cresset Press, London.

Davis, E. B. (1994). 'Parcere nominibus': Boyle, Hooke and the rhetorical interpretation of Descartes. In Hunter 1994b, 157–175.

Dear, Peter (2001). *Revolutionizing the Sciences: European Knowledge and its Ambitions, 1500–1700*. Palgrave, Basingstoke.

De Gant, François (1995). *Force and Geometry in Newton's* Principia, trans. Curtis Wilson. Princeton University Press, Princeton.

Dennis, M. A. (1989). Graphic understanding: instruments and interpretation in Robert Hooke's *Micrographia*. *Science in Context* 3: 309–364.

Drake, E. T. (1996). *Restless Genius. Robert Hooke and his Earthly Thoughts*. Oxford University Press, New York.

Ehrlich, M. E. (1995). Mechanism and activity in the Scientific Revolution: the case of Robert Hooke. *Annals of Science* 52: 127–151.

'Espinasse, Margaret (1956). *Robert Hooke*. Heinemann, London (reprint, University of California Press, Berkeley and Los Angeles, 1962).

Evelyn, John (1955). *Diary*, ed. E. S. de Beer. 6 vols. Clarendon Press, Oxford.

Featherstone, Ernest (1952). *Sir Thomas Gresham and his Trusts*. Blades, East & Blades, London.

Feingold, Mordechai (2000). Mathematicians and naturalists: Sir Isaac Newton and the Royal Society. In J. Z.Buchwald and I. B. Cohen (eds.), *Isaac Newton's Natural Philosophy*. MIT Press, Cambridge, Mass., 77–102.

Flamsteed, John (1975). *The Gresham Lectures*, ed. E. G. Forbes. Mansell, London.

Frank, R. G. (1980). *Harvey and the Oxford Physiologists. Scientific Ideas and Social Interraction*. University of California Press, Berkeley and Los Angeles.

Fulton, J. F. (1961). *A Bibliography of the Honourable Robert Boyle F.R.S.* 2nd ed. Clarendon Press, Oxford.

Gal, Ofer (1996). Producing knowledge in the workshop: Hooke's 'inflection' from optics to planetary motion. *Studies in History and Philosophy of Science* 27: 181–205.

Gaukroger, Stephen (2001). *Francis Bacon and the Transformation of Early-Modern Philosophy*. Cambridge University Press.

Geraghty, Anthony (2001). Edward Woodroofe: Sir Christopher Wren's first draughtsman. *Burlington Magazine* 143: 474–479.

Gouk, Penelope (1980). The role of acoustics and music theory in the scientific work of Robert Hooke. *Annals of Science* 37: 573–605.

Gouk, Penelope (1999). *Music, Science and Natural Magic in Seventeenth-Century England*. Yale University Press, New Haven and London.

Gunther, R. T. (ed.) (1930a). *Early Science in Oxford*. Vol. 6. *The Life and Work of Robert Hooke (Part I)*. Oxford.

Gunther, R. T. (ed.) (1930b). *Early Science in Oxford*. Vol. 7. *The Life and Work of Robert Hooke (Part II)*. Oxford.

Gunther R. T. (ed.) (1935). *Early Science in Oxford*. Vol. 10. *The Life and Work of Robert Hooke (Part IV)*. Oxford.

Hall, A. R. (1951). Two unpublished lectures of Robert Hooke. *Isis* 42: 219–230.

Hall, A. R. (1966). *Hooke's Micrographia 1665–1965*. Athlone Press, London.

Hall, A. R. (1978). Horology and criticism: Robert Hooke. *Studia Copernicana* 16: 261–281.

Hall, A. R., and Hall, M. B. (eds.) (1962). *Unpublished Scientific Papers of Isaac Newton*. Cambridge University Press, Cambridge.

Halley, Edmond (1937). *Correspondence and Papers*, ed. E. F. MacPike. Taylor and Francis, London.

Hambly, E. C. (1987). Robert Hooke, the City's Leonardo. *City University* 2(2): 5–10.

Harkness, D. E. (1999). *John Dee's Conversations with Angels: Cabala, Alchemy and the End of Nature*. Cambridge University Press, Cambridge.

Hartley, Sir Harold (ed.) (1960). *The Royal Society: Its Origins and Founders*. The Royal Society, London.

Harwood, John (1989). Rhetoric and graphics in *Micrographia*. In Hunter and Schaffer 1989, 119–147.

Henry, John (1986). Occult qualities and the experimental philosophy: active principles in pre-Newtonian matter theory. *History of Science* 24: 335–381.

Henry, John (1989). Robert Hooke, the incongruous mechanist. In Hunter and Schaffer 1989, 149–180.

Hesse, Mary (1966). Hooke's philosophical algebra. *Isis* 57: 67–83.

Hooke, Robert (1661). *An Attempt for the Explication of the Phænomena, Observable in an Experiment Published by the Honourable Robert Boyle, Esq; . . . In Confirmation of a former Conjecture made by R. H.* London.

Hooke, Robert (1665). *Micrographia: Or Some Physiological Descriptions of Minute Bodies Made By Magnifying Glasses With Observations and Enquiries thereupon*. London (reprinted in *Early Science in Oxford*, vol. 13 (Oxford, 1938), and by Dover Books, New York, 1961).

Hooke, Robert (1674). *Animadversions On the first part of the Machina Cœlestis of. . . . Johannes Hevelius . . .* London.

Hooke, Robert (1679). *Lectiones Cutlerianæ*. London (reprint with new pagination as *Early Science in Oxford*. Vol. 8. *The Cutler Lectures of Robert Hooke*, ed. R. T. Gunther, Oxford, 1931).

Hooke, Robert (1705). *Posthumous Works*, ed. Richard Waller. London (reprints, Johnson Reprint Corporation, New York, 1969; Frank Cass, London, 1971).

Hooke, Robert (1726). *Philosophical Experiments and Observations of the late Eminent Dr. Robert Hooke* . . . ed. William Derham. London (reprint, Frank Cass, London, 1962).

Hooke, Robert (1935). *The Diary of Robert Hooke 1672–1680*, ed. H. W. Robinson and Walter Adams. Taylor & Francis, London.

Hooke, Robert (1989). *Micrografía o Algunas Descripciones Fisiológicas de Los Cuerpos Diminutos Realizadas Mediante Christales de Aumento con Observaciones y Disquisiciones sobre Ellas*, ed. Carlos Solís. Alfaguara, Madrid.

Howgego, J. L. (1978). *Printed Maps of London circa 1553–1850*. 2nd ed. Dawson, Folkestone.

Howse, Derek (1970–1). The Tompion clocks at Greenwich and the dead-beat escapement. *Antiquarian Horology* 7: 18–34, 114–133.

Howse, Derek (1975). *Greenwich Observatory*. Vol. 3. *The Buildings and Instruments*. Taylor and Francis, London.

Hull, Derek (1997). Robert Hooke: a fractographic study of Kettering-stone. *Notes and Records of the Royal Society of London* 51: 45–55.

Hunter, Michael (1975). *John Aubrey and the Realm of Learning*. Gerald Duckworth, London.

Hunter, Michael (1984). A 'college' for the Royal Society: the abortive plan of 1667–8. *Notes and Records of the Royal Society* 38: 159–186.

Hunter, Michael (1989) *Establishing the New Science: The Experience of the Early Royal Society*. Boydell Press, Woodbridge.

Hunter, Michael (1994a). *Robert Boyle by Himself and His Friends*. Pickering & Chatto, London.

Hunter, Michael (ed.) (1994b). *Robert Boyle Reconsidered*. Cambridge University Press.

Hunter, Michael (1994c). *The Royal Society and its Fellows 1660–1700: The Morphology of an Early Scientific Institution*. Rev. ed. British Society for the History of Science, Oxford.

Hunter, Michael (2000). *Robert Boyle (1627–91): Scrupulosity and Science*. Boydell Press, Woodbridge.

Hunter, Michael, and Schaffer, Simon (eds.) (1989). *Robert Hooke: New Studies*. Boydell Press, Woodbridge.

Iliffe, Rob (1992). 'In the warehouse': privacy, property and priority in the early Royal Society. *History of Science* 30: 29–68.

Iliffe, Rob (1995). Material doubts: Hooke, artisan culture and the exchange of information in 1670s London. *British Journal for the History of Science* 28: 285–318.

Ito, Yushi (1988). Hooke's cyclic theory of the earth in the context of seventeenth-century England. *British Journal for the History of Science* 21: 295–314.

Jardine, Lisa (1999). *Ingenious Pursuits: Building the Scientific Revolution.* Little, Brown, London.

Jardine, Lisa (2002). *On a Grander Scale: The Outstanding Career of Sir Christopher Wren.* HarperCollins, London.

Jardine, Lisa, and Stewart, Alan (1998). *Hostage to Fortune: The Troubled Life of Francis Bacon.* Gollancz, London.

Jeffery, Paul (1996). *The City Churches of Sir Christopher Wren.* Hambledon Press, London.

Johns, Adrian (1998). *The Nature of the Book. Print and Knowledge in the Making.* Chicago University Press, Chicago.

Jones, P. E. (ed.) (1966). *The Fire Court.* Vol. 1. *Calendar to the Judgements and Decrees of the Court of Judicature appointed to determine differences between landlords and tenants as to rebuilding after the Great Fire.* William Clowes, London.

Jones, P. E. and Reddaway, T. F. (1967). Introduction. In London Topographical Society 1967, ix–xlii.

Kassler, J. C. (1995). *Inner Music: Hobbes, Hooke and North on Internal Character.* Athlone, London.

Kassler, J. C., and Oldroyd, David (1983). Robert Hooke's Trinity College 'Musick Scripts': his music theory and the role of music in his cosmology. *Annals of Science* 40: 559–595.

Kent, Paul (2001). *Some Scientists in the Life of Christ Church, Oxford.* Oxford University Press.

Keynes, Sir Geoffrey (1960). *A Bibliography of Dr Robert Hooke.* Clarendon Press, Oxford.

Koyré, Alexandre (1952). An unpublished letter of Robert Hooke to Isaac Newton. *Isis* 43: 312–337. Reprinted in Koyré's *Newtonian Studies.* Chicago University Press, Chicago, 1965, 221–60.

Lang, Jane (1956). *Rebuilding St. Paul's after the Great Fire of London.* Oxford University Press.

LeFanu, William (1990). *Nehemiah Grew A Study and Bibliography of his Writings.* St Paul's Bibliographies, Winchester.

Lohne, J. A. (1960). Hooke versus Newton: an analysis of the documents in the case on free fall and planetary motion. *Centaurus* 7: 6–52.

London Topographical Society (1962a, b, c). *The Survey of Building Sites in the City of London after the Great Fire of 1666.* Vols. 3, 4, and 5. London Topographical Society, London.

London Topographical Society (1964, 1967). *The Survey of Building Sites in the City of London after the Great Fire of 1666.* Vols. 2 and 1. London Topographical Society, London.

Love, Harold (1998). *The Culture and Commerce of Texts: Scribal Publication in Seventeenth-century England.* 2nd ed. University of Massachusetts Press, Amherst, Mass.

Lynch, W. T. (2001). *Solomon's Child. Method in the Early Royal Society of London.* Stanford University Press, Stanford, Calif.

Lyons, Sir Henry (1944). *The Royal Society 1660–1940.* Cambridge University Press.

McConnell, Anita (1982). *No Sea Too Deep: A History of Oceanographic Instruments.* Adam Hilger, Bristol.

McKie, D. (1953). Flame and the *flamma vitalis*: Boyle, Hooke and Mayow. In E. A. Underwood (ed.), *Science, Medicine and History.* 2 vols. Oxford University Press, London, i, 469–488.

Maddison, R. E. W. (1969). *The Life of the Honourable Robert Boyle F. R. S.* Taylor & Francis, London.

Masters, B. R. (1974). *Leybourn's Plans of London Markets, 1677.* London Topographical Society, London.

Middleton, W. E. K. (1969). *The History of the Barometer.* Johns Hopkins University Press, Baltimore.

Milne, Gustav (1986). *The Great Fire of London.* Historical Publications, New Barnet.

Mulligan, Lotte (1992). Robert Hooke's 'Memoranda': memory and natural history. *Annals of Science* 49: 47–61.

Mulligan, Lotte (1996). Self scrutiny and the study of nature: Robert Hooke's diary as natural history. *Journal of British Studies* 35: 311–342.

Nakajima, Hideto (1994). Robert Hooke's family and his youth: some new evidence from the will of the Rev. John Hooke. *Notes and Records of the Royal Society of London* 48: 11–16.

Nauenberg, Michael (1994). Hooke, orbital motion, and Newton's Principia. *American Journal of Physics* 62: 331–50.

Newton, Isaac (1959–77). *Correspondence*, ed. H. W. Turnbull, J. F. Scott, A. R. Hall, and Laura Tilling. 7 vols. Cambridge University Press.

Oldenburg, Henry (1965–86). *Correspondence*, ed. A. R. and M. B. Hall. 13 vols. University of Wisconsin Press, Madison; Mansell, London; Taylor & Francis, London.

Oldroyd, David (1972). Robert Hooke's methodology of science as exemplified in his 'Discourse of Earthquakes'. *British Journal for the History of Science* 6: 109–130.

Oldroyd, David (1980). Some 'Philosophicall Scribbles' attributed to Robert Hooke. *Notes and Records of the Royal Society of London* 35: 17–32.

Oldroyd, David (1987). Some writings of Robert Hooke on procedures for the prosecution of scientific inquiry, including his 'Lectures of Things Requisite to a Natural History'. *Notes and Records of the Royal Society* 41: 145–167.

Oldroyd, David (1989). Geological controversy in the seventeenth century: Hooke vs. Wallis and its aftermath. In Hunter and Schaffer 1989, 207–233.

Patterson, L. D. (1949–50). Hooke's gravitation theory and its influence on Newton. *Isis* 40: 327–341, 41: 32–45.

Pepys, Samuel (1970–83). *Diary*, ed. R. C. Latham and William Matthews. 11 vols. Bell & Hyman, London.

Perks, Sydney (1922). *The History of the Mansion House*. Cambridge University Press, Cambridge.

Porter, Roy, and Rousseau, G. S. (1998). *Gout: The Patrician Malady*. Yale University Press, New Haven and London.

Porter, Stephen (1996). *The Great Fire of London*. Sutton, Stroud.

Pugliese, P. J. (1982). *The Scientific Achievement of Robert Hooke*. PhD thesis, Harvard University. University Microfilms International, Ann Arbor. Michigan, and London.

Pugliese, P. J. (1989). Robert Hooke and the dynamics of motion in a curved path. In Hunter and Schaffer 1989, 181–205.

Pumfrey, Stephen (1991). Ideas above his station: a social study of Hooke's curatorship of experiments. *History of Science* 29: 1–44.

Rappaport, Rhoda (1986). Hooke on earthquakes: lectures, strategy and audience. *British Journal for the History of Science* 19: 129–146.

Rappaport, Rhoda (1997). *When Geologists were Historians, 1665–1750*. Cornell University Press, Ithaca, NY.

Rasmussen, S. E. (1937). *London: The Unique City*. Penguin Books, Harmondsworth (abridged edition, 1960).

Reddaway, T. F. (1940). *The Rebuilding of London after the Great Fire*. Jonathan Cape, London (reprint with minor changes, Edward Arnold, London, 1951).

Sabra, A. I. (1967). *Theories of Light from Descartes to Newton*. Oldbourne, London.

Sargeaunt, John (1898). *Annals of Westminster School*. Methuen, London.

Saunders, Ann (1997). The second Exchange 1669–1838. In Ann Saunders (ed.) (1997), *The Royal Exchange*. London Topographical Society, London, 121–135.

Shadwell, Thomas (1966). *The Virtuoso*, ed. M. H. Nicolson and David Rodes. University of Nebraska Press, Lincoln and London.

Shapin, Steven (1988a). The house of experiment in seventeenth-century England. *Isis* 79: 373–404.

Shapin, Steven (1988b). Robert Boyle and mathematics: reality, representation and experimental practice. *Science in Context* 2: 23–58.

Shapin, Steven (1989). Who was Robert Hooke? In Hunter and Schaffer 1989, 253–285.

Shapin, Steven (1994). *A Social History of Truth*. Chicago University Press.

Shapin, Steven, and Schaffer, Simon (1985). *Leviathan and the Air-pump. Hobbes, Boyle, and the Experimental Life*. Princeton University Press, Princeton.

Simpson, A. D. C. (1989). Robert Hooke and practical optics: technical support at a scientific frontier. In Hunter and Schaffer 1989, 33–61.

Singer, B. R. (1976). Robert Hooke on memory, association and perception. *Notes and Records of the Royal Society* 31: 115–131.

Sprat, Thomas (1667). *The History of the Royal Society*. London.

Stoesser-Johnston, Alison (1997). *Robert Hooke and Holland: Dutch Influence on Hooke's Architecture*. Doctoraalsscripties Bouwkunst. Rijksuniversiteit Utrecht.

Taylor, E. G. R. (1937). Robert Hooke and the cartographical projects of the late seventeenth century (1666–1696). *Geographical Journal* 90: 529–540.

Taylor, E. G. R. (1940). 'The English Atlas' of Moses Pitt, 1680–83. *Geographical Journal* 95: 292–299.

Turner, Anthony (1974). Hooke's theory of the earth's axial displacement: some contemporary opinion. *British Journal for the History of Science* 7: 166–170.

Turner, H. D. (1956). Robert Hooke and theories of combustion. *Centaurus* 4: 297–310.

Walker, D. P. (1958). *Spiritual and Demonic Magic from Ficino to Campanella*. Studies of the Warburg Institute, vol. 22. Warburg Institute, London (reprint, University of Notre Dame Press, Notre Dame, Ind., 1975).

Ward, John (1740). *Lives of the Professors of Gresham College*. London.

Ward, Ned (1927). *The London Spy*, ed. A. L. Hayward. Cassell, London.

Webster, Charles (1963). Richard Towneley and Boyle's Law. *Nature* 197: 226–228.

Webster, Charles (1965). The discovery of Boyle's Law, and the concept of the elasticity of air in the seventeenth century. *Archive for History of Exact Sciences* 2: 441–502.

Westfall, R. S. (1969). Introduction to the Johnson Reprint Corporation reprint of Hooke 1705. New York.

Westfall, R. S. (1972). Hooke, Robert. In *Dictionary of Scientific Biography*, ed. C. C. Gillispie, Charles Scribner's Sons, New York, 6: 481–488.

Westfall, R. S. (1980). *Never at Rest. A Biography of Isaac Newton*. Cambridge University Press, Cambridge.

Westfall, R. S. (1983). Robert Hooke, mechanical technology and scientific investigation. In J. G. Burke (ed.), *The Uses of Science in the Age of Newton*. University of California Press, Berkeley and Los Angeles, 85–110.

Westman, R. S. (1980). The astronomer's role in the sixteenth century: a preliminary study. *History of Science* 18: 105–147.

Wood, P. B. (1980). Methodology and apologetics: Thomas Sprat's *History of the Royal Society*. *British Journal for the History of Science* 13: 1–26.

Woodhead, J. R. (1965). *The Rulers of London 1660–1689*. London & Middlesex Archaeological Society, London.

Wren Society (1924–43). Wren Society Publications. 20 vols. Oxford University Press, Oxford.

Wright, Michael (1989). Robert Hooke's longitude timekeeper. In Hunter and Schaffer 1989, 63–118.

INDEX

[illegible] rose at 6. sent to Scarborough [illegible] to Guildhall about works [illegible] for Crisp. [illegible]
Andrew King. [illegible] from Dr. Mason [illegible] Polyppe [illegible] little berry [illegible] at Garaways [illegible] Harry [illegible]
[illegible] of pendant watch. of spring watches [illegible] Dr. Mayow [illegible] bath water [illegible]
[illegible] Bedlam [illegible] Crisp to my dining room [illegible] chocolatt. [illegible] Hill [illegible] stone pottery. Colery [illegible]
[illegible] Garaways. drank 2 dish chocolatt [illegible] Scarborough
[illegible] spoke about [illegible] at Fleetstreet.

[illegible] of ridiculous new watch [illegible]
14. at Garaways about [illegible] 13 [illegible] Sir Chr. Wren [illegible] to Sir Rich. Ford [illegible] Sir Jonas Moore [illegible]
6. at Spanish Coffee house [illegible] Hill [illegible] Lodowick discourse of [illegible] Language. of ye temporall [illegible]
Garaways [illegible] Jenkins [illegible] Hammond [illegible] Birchin Lane [illegible] sack [illegible] melancholy
[illegible] 1 pr of Black silk stockins [illegible] black silk stock [illegible]
[illegible] at Garaways. [illegible] Bowen [illegible] market [illegible] 1 pm. DH. [illegible] velvet
[illegible] 9½ pm. Libera me Domine. [illegible] Aubery [illegible] till 11½ at night discoursed to him my designe about [illegible]
principle about flying [illegible] he told me of Dr. Platts inquiry for a [illegible]
[illegible] at Bedlam [illegible] Sir R. Piggot [illegible]
[illegible] Garaways. At [illegible] about fitting [illegible] 4 p m. DH. pullet. [illegible]
[illegible] chocolat [illegible] Bedlam [illegible] Hayward about Roof
[illegible] at Garaways [illegible] 2 dish chocolat [illegible] Barrington Hill [illegible] A description of urnes found under an old
[illegible] Dr. Sydenhams brother. wrote to Brother. [illegible]
[illegible] good [illegible] malago. [illegible] black [illegible]
[illegible] to Guildhall [illegible] account [illegible] at Garaways. at the [illegible] head in [illegible] found [illegible]
[illegible] Scarborough [illegible] Old bayly [illegible] Glisson Scarborough Whistler [illegible]
[illegible] the Colledge Seale and Deliver the bond for 100 lb [illegible]
[illegible] painted by Wagoner at [illegible] afterwards that [illegible]
[illegible] speculum metall. [illegible] plates of french glasse. Delivered [illegible]
[illegible] Sir Ch. Wren. [illegible] at Garaways [illegible] Hill [illegible] Andrews. Scarborough
[illegible] about [illegible]
December. 18. [illegible] smoked 2 p. at 8 hour. [illegible] at [illegible]
to [illegible] Fitch [illegible] black [illegible] 60 double plates [illegible] to Sir C.
[illegible] Garaways discourse with Capt Sheers
[illegible] Mercers chapp [illegible] magneticall [illegible] body of ye earth. [illegible]
[illegible] about ships [illegible] Hammond bills [illegible] Berlen his [illegible]
[illegible] DH. [illegible] Hammond [illegible] about [illegible] Creation
[illegible] Garaways [illegible] Hill [illegible] drank port
[illegible] all vegetables to be [illegible] Aubery [illegible] flying [illegible] drank port [illegible]
Dec. 19. Slept [illegible] rose at 11. [illegible] Harry [illegible] Whistler. at [illegible] drank port [illegible] smoked 4 p. [illegible]
[illegible] Garaways [illegible] Tompion to Sir J. Moore [illegible] sent [illegible]
[illegible] Tompion. [illegible] drank tea [illegible] to Sir Ch. Wr. he told me of ye grand designe. [illegible]
[illegible] ye Chamber. to Com[illegible] for [illegible] to little brittain. Blackburne at little [illegible] Hammond. DH. at [illegible] King [illegible]
[illegible] Whistler. [illegible] Lane Coffee house. Garaways [illegible] Hammond [illegible]
[illegible] till 12½ p.m. The Com[illegible] for [illegible] ordered me 100 lb for last years attendance.
[illegible] Hammond to [illegible] Whistler. who undertook to lend Mr Hammond 300 lb [illegible] 200
[illegible] 21. [illegible] Hammond [illegible] at Garaways discoursed about Holder [illegible]
Colledge. Mr Hammond [illegible] at Garaways little room [illegible] Hill Hopegood [illegible]
[illegible] about [illegible] resolved to give him a microscope. at Mr. Hills [illegible]
[illegible] Hook [illegible] Garaways [illegible] about [illegible]
[illegible] not to bed but saw Eclipse observed [illegible]
22. Rose early [illegible] generall [illegible] paid me 7. 10 for 1 quarter [illegible]
[illegible] 100 lb [illegible] which made [illegible] besides [illegible] from Montac. [illegible]
[illegible] from Sir John [illegible] 19 [illegible] for Coach. At Garaways [illegible] DH. [illegible]
[illegible] about [illegible] much discourse [illegible] Gave Hook at Gar[illegible]
Chocolatt. [illegible] a man had been [illegible] Pike. [illegible]
[illegible] drank chocolat [illegible] port [illegible] the reflective proper[illegible]
[illegible] like Kings [illegible] glasses [illegible]
23. to Garaways early [illegible] 12 small dishes 6 lb for 2 great 6 lb. [illegible] chocolat. [illegible] took at Coac[illegible]
[illegible] watch in pocket. [illegible] Hills discoursed of patent [illegible] other things. called on Tompion [illegible]
Sir Ch: Wren [illegible] in Old bayly. DH. [illegible] velvet coat [illegible] 10 lb [illegible]
[illegible] with Mr Hill at [illegible] drank malago made me very sick. Mr Lodowick discoursed about
[illegible] about flying [illegible] chocolatt. Slept well. Hammond [illegible] appointed to morrow [illegible]
24. at Garaways [illegible] Whistler [illegible] Hammond Jenkins wrot note of things to be done at the Colledge. [illegible] Hammond [illegible]
[illegible] to Controuler. Smoked & drank [illegible] Sir R. Ford. [illegible] warrant from Controuler for 100 lb [illegible] Hook. Hill [illegible]
[illegible] new book [illegible] Kings anatom: [illegible] Crisp drank portport. Dind with Dean Tillotson. Gave [illegible]
[illegible] watch [illegible] 50 lb [illegible] of 4000 [illegible] a small Roma subterranea. [illegible]
Garaways. Andrews. about paper. about [illegible] lane houses [illegible] Newbold. Scarborough [illegible]
[illegible] Chamber. took [illegible]

[illegible]